T0204000

Applied Regularization Methods for the Social Sciences

Chapman & Hall/CRC
Statistics in the Social and Behavioral Sciences Series

Series Editors

Jeff Gill, Steven Heeringa, Wim J. van der Linden, Tom Snijders

Recently Published Titles

Multilevel Modeling Using R, Second Edition
W. Holmes Finch, Joselyn E. Bolin, and Ken Kelley

Modelling Spatial and Spatial-Temporal Data: A Bayesian Approach
Robert Haining and Guangquan Li

Handbook of Automated Scoring: Theory into Practice
Duanli Yan, André A. Rupp, and Peter W. Foltz

Interviewer Effects from a Total Survey Error Perspective
Kristen Olson, Jolene D. Smyth, Jennifer Dykema, Allyson Holbrook, Frauke Kreuter, and Brady T. West

Measurement Models for Psychological Attributes
Klaas Sijtsma and Andries van der Ark

Big Data and Social Science: Data Science Methods and Tools for Research and Practice, Second Edition
Ian Foster, Rayid Ghani, Ron S. Jarmin, Frauke Kreuter and Julia Lane

Understanding Elections through Statistics: Polling, Prediction, and Testing
Ole J. Forsberg

Analyzing Spatial Models of Choice and Judgment, Second Edition
David A. Armstrong II, Ryan Bakker, Royce Carroll, Christopher Hare, Keith T. Poole and Howard Rosenthal

Introduction to R for Social Scientists: A Tidy Programming Approach
Ryan Kennedy and Philip Waggoner

Linear Regression Models: Applications in R
John P. Hoffman

Mixed-Mode Surveys: Design and Analysis
Jan van den Brakel, Bart Buelens, Madelon Cremers, Annemieke Luiten, Vivian Meertens, Barry Schouten and Rachel Vis-Visschers

Applied Regularization Methods for the Social Sciences
Holmes Finch

For more information about this series, please visit: www.routledge.com/Chapman-HallCRC-Statistics-in-the-Social-and-Behavioral-Sciences/book-series/CHSTSOBESCI

Applied Regularization Methods for the Social Sciences

Holmes Finch

CRC Press
Taylor & Francis Group
Boca Raton London New York

CRC Press is an imprint of the
Taylor & Francis Group, an **informa** business

A CHAPMAN & HALL BOOK

First edition published 2022
by CRC Press
6000 Broken Sound Parkway NW, Suite 300, Boca Raton, FL 33487–2742

and by CRC Press
2 Park Square, Milton Park, Abingdon, Oxon, OX14 4RN

CRC Press is an imprint of Taylor & Francis Group, LLC

Library of Congress Cataloging-in-Publication Data
Names: Finch, W. Holmes (William Holmes), author.
Title: Applied regularization methods for the social sciences / Holmes Finch.
Description: First edition. | Boca Raton : CRC Press, 2022. | Includes bibliographical references and index.
Identifiers: LCCN 2021048064 (print) | LCCN 2021048065 (ebook) |
   ISBN 9780367408787 (hardback) | ISBN 9781032209470 (paperback) |
   ISBN 9780367809645 (ebook)
Subjects: LCSH: Social sciences—Statistical methods. | Big data. | Mathematical
   statistics. | R (Computer program language)
Classification: LCC HA29 .F475 2022 (print) | LCC HA29 (ebook) |
   DDC 300.1/5—dc23/eng/20220103
LC record available at https://lccn.loc.gov/2021048064
LC ebook record available at https://lccn.loc.gov/2021048065

ISBN: 978-0-367-40878-7 (hbk)
ISBN: 978-1-032-20947-0 (pbk)
ISBN: 978-0-367-80964-5 (ebk)

DOI: 10.1201/9780367809645

Typeset in Palatino
by Apex CoVantage, LLC

Contents

1

R

The goal of this chapter is to provide a simple introduction to R for those readers not familiar with it. In addition, readers familiar with R but who would like a brief review of basic issues regarding data handling, variable creation, and the like will hopefully find this chapter to be helpful. R is what is known as open source software, meaning that anyone is able to write functions and add to the software. The process for doing this is governed by a nonprofit organization housed in Vienna, Austria, and more information can be found at http://cran.us.r-project.org/. In addition, R is freely available to users on a wide variety of computing platforms, most notably Windows, Macintosh OS, and Unix/Linux. Thus, R offers researchers a potentially valuable tool for conducting statistical analyses ranging from the most basic to complex, cutting-edge methods. It is important to note that because R is open source, developers do update their functions fairly frequently, occasionally rendering commands that worked at one time inoperable. Therefore, you will need to keep the functions you use up to date to ensure that you maximize their functionality and that they continue to work appropriately for you. In the following sections, we will review the basic elements of R and how data analysts interact with it. One final point here is that we will be focusing on the most stripped-down, basic form of R. There are other interfaces for interacting with the software, in particular, Rstudio. For more information on Rstudio, you are encouraged to go to www.rstudio.com/.

The R Console and R Scripts

When working with R, commands are entered at the red > prompt, after which you will hit the return key to execute it. In Figure 1.1, we can see an image of the basic R console. If we type in a command and would like to do it again, we can simply use the up and down arrow keys to cycle through all previous commands and find the one that we would like.

Using the console in this fashion can be very slow and laborious if we have many commands to carry out at once. In addition, the commands are not saved in a convenient locale. Therefore, we will find it easier to type our commands and function calls into the script editor that is a part of R. We can

DOI: 10.1201/9780367809645-1

FIGURE 1.1
R console

open the script editor using the menu sequence: FILE>NEW SCRIPT, at which point an empty script window will open up. We can then type commands in this window that we would like to submit to R and then submit code to run by selecting the commands that we would like to run and clicking ▐◀▐ [on the menu bar. The code is automatically submitted to the R console. As an example, consider the basic script below in which we create a variable called `test.me` consisting of 5 numbers and then request its mean.

```
test.me<-c(5,4,7,4,9,6,10,1)
mean(test.me)
```

In order to carry these commands out, we highlight them in R and then click the submit icon displayed earlier. The R console then appears as below in Figure 1.2.

FIGURE 1.2
Obtaining mean in R

R Libraries

The main base R software contains many of the functions that we will need for conducting statistical analyses and producing graphical results. However, many more specialized techniques, including most of those used in this text, require that we download and install libraries specially designed for that purpose. In order to install these additional packages, we would simply use the following menu command sequence: PACKAGES>INSTALL PACKAGES.

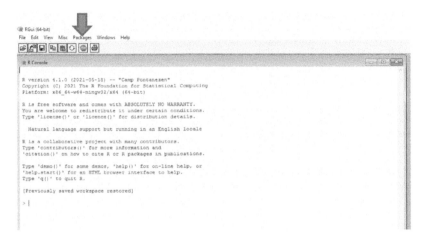

FIGURE 1.3
Installing packages in R

We may be asked to select a "mirror", which is simply a computer from which we download the package of interest. Generally speaking, any of these mirrors will be fine. Next, we then see the following window:

> We would then need to select a server (morror) from which to download the libraries. Any of them should be equally fine for this purpose.

FIGURE 1.4
CRAN mirror

Once we select the mirror, we are then presented with a list of all available libraries from which we can select the one(s) that we want to install.

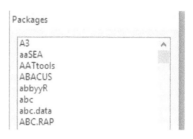

FIGURE 1.5
R packages

Let's install the glmnet library, which we will be using throughout this text. We would scroll through the list and select the package that we would like to download.

FIGURE 1.6
GLMNET package

After clicking OK, the package will automatically download to our computer and be installed in the form of a library. Once a library is installed, we can access it simply by typing the `library` command. For example, the command `library(glmnet)` would install the `glmnet` library that we will use to conduct many of the SEM analyses in the following chapter. Once the library has been installed, we can then load it using the `library` command, as in `library(glmnet)`.

Reading and Viewing Data in R

R allows us to read data from a variety of sources, including SPSS, Excel, or .txt files to name a few. We can read .txt files using the `read.table` command.

If the data file has variable names in the first row, we would use the `header=T` statement to indicate that the top row of values in the file are variable names. For example, if we wish to open a text file in the c:\ directory called test.txt, which has the variable names in the first row, we could use the following command:

```
test <- read.table(c:\test.txt, header=T)
```

We can read SPSS data files using the `read.spss` function in the foreign library, and Excel files using the `read _ excel` function from the `readxl` library. We will see examples of doing this throughout the book, with the code available in the scripts at www.routledge.com/9780367408787. Finally, when the data are read into R it is saved in what is known as a dataframe, which corresponds to a data file in SPSS, SAS, or Stata.

Once the data are read into R, we can view it using the View command. For example, if we have a dataframe called full_data, we can open it using the following command:

```
View(full_data)
```

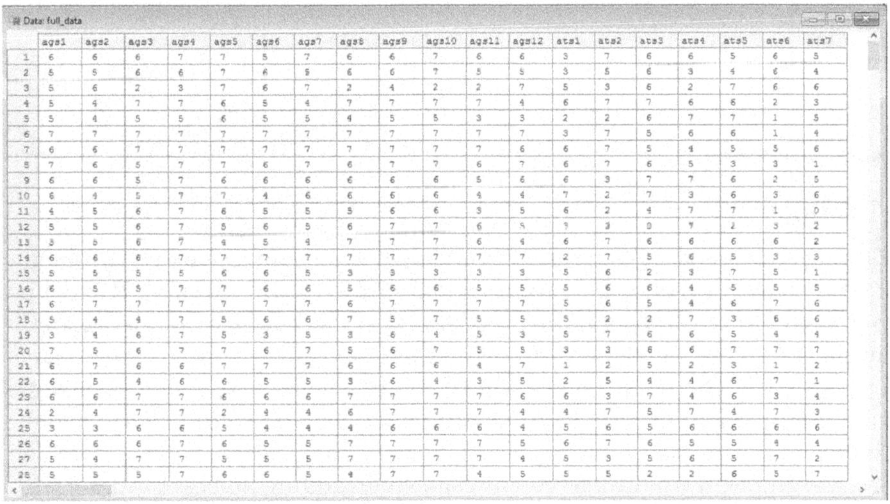

Data: full_data

	ags1	ags2	ags3	ags4	ags5	ags6	ags7	ags8	ags9	ags10	ags11	ags12	ats1	ats2	ats3	ats4	ats5	ats6	ats7
1	6	6	6	7	7	5	7	6	6	7	6	6	3	7	6	6	5	6	5
2	5	5	6	6	7	6	5	6	6	7	5	5	3	5	6	3	4	6	4
3	5	6	2	3	7	6	7	2	4	2	2	7	5	3	6	2	7	6	6
4	5	4	7	7	6	5	4	7	7	7	7	4	6	7	7	6	6	2	3
5	5	4	5	5	6	5	5	4	5	5	3	3	2	2	6	7	7	1	5
6	7	7	7	7	7	7	7	7	7	7	7	7	3	7	5	6	6	1	4
7	6	6	7	7	7	7	7	7	7	7	7	6	6	7	5	4	5	5	6
8	7	6	5	7	7	6	7	6	7	7	6	7	6	7	6	5	3	3	1
9	6	6	5	7	6	6	6	6	6	6	5	6	6	3	7	7	6	2	5
10	6	4	5	7	7	4	6	6	6	6	4	4	7	2	7	3	6	3	6
11	4	5	6	7	6	5	5	5	6	6	3	5	6	2	4	7	7	1	0
12	5	5	6	7	5	6	5	6	7	7	6	5	[illegible]	3	0	7	2	3	2
13	3	5	6	7	4	5	4	7	7	7	6	4	6	7	6	6	6	6	2
14	6	6	6	7	7	7	7	7	7	7	7	7	2	7	5	6	5	3	3
15	5	5	5	5	6	6	5	3	3	3	3	3	5	6	2	3	7	5	1
16	6	5	5	7	7	6	6	5	6	6	5	5	5	6	6	4	5	5	5
17	6	7	7	7	7	7	7	6	7	7	7	7	5	6	5	4	6	7	6
18	5	4	4	7	5	6	6	7	5	7	5	5	5	2	2	7	3	6	6
19	3	4	6	7	5	3	5	3	6	4	5	3	5	7	6	6	5	4	4
20	7	5	6	7	7	6	7	5	6	7	5	5	3	3	6	6	7	7	7
21	6	7	6	6	7	7	7	6	6	6	4	7	1	2	5	2	3	1	2
22	6	5	4	6	6	5	5	3	6	4	3	5	2	5	4	4	6	7	1
23	6	6	7	7	6	6	6	7	7	7	7	6	6	3	7	4	6	3	4
24	2	4	7	7	2	4	4	6	7	7	7	4	4	7	5	7	4	7	3
25	3	3	6	6	5	4	4	4	6	6	6	4	5	6	5	6	6	6	6
26	6	6	6	7	6	5	5	7	7	7	7	5	6	7	6	5	5	4	4
27	5	4	7	7	5	5	5	7	7	7	7	4	5	3	5	6	5	7	2
28	5	5	5	7	6	6	5	4	7	7	4	5	5	5	2	2	6	5	7

FIGURE 1.7
Data view

If we would like to edit the data we would use the `edit(full _ data)` and obtain the following window, in which we can edit the data values.

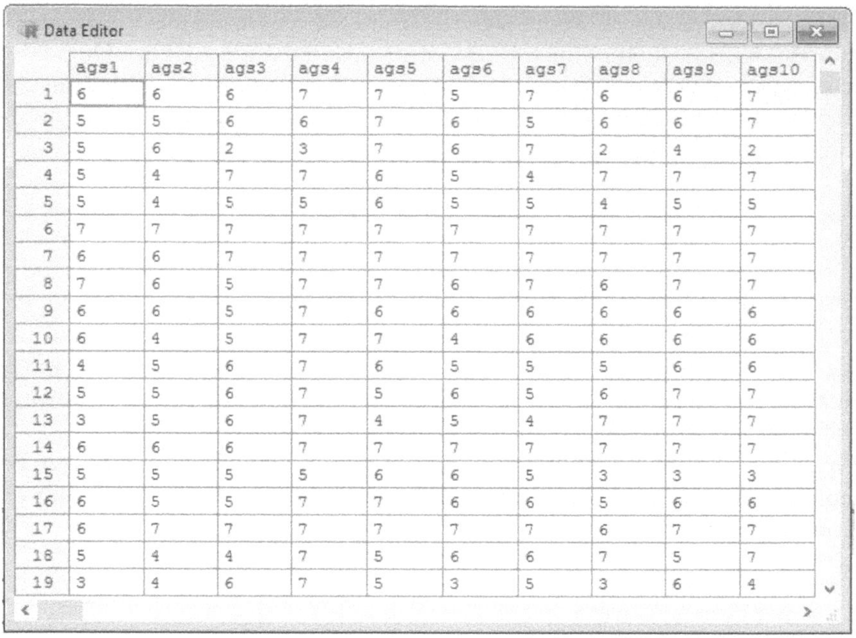

	ags1	ags2	ags3	ags4	ags5	ags6	ags7	ags8	ags9	ags10
1	6	6	6	7	7	5	7	6	6	7
2	5	5	6	6	7	6	5	6	6	7
3	5	6	2	3	7	6	7	2	4	2
4	5	4	7	7	6	5	4	7	7	7
5	5	4	5	5	6	5	5	4	5	5
6	7	7	7	7	7	7	7	7	7	7
7	6	6	7	7	7	7	7	7	7	7
8	7	6	5	7	7	6	7	6	7	7
9	6	6	5	7	6	6	6	6	6	6
10	6	4	5	7	7	4	6	6	6	6
11	4	5	6	7	6	5	5	5	6	6
12	5	5	6	7	5	6	5	6	7	7
13	3	5	6	7	4	5	4	7	7	7
14	6	6	6	7	7	7	7	7	7	7
15	5	5	5	5	6	6	5	3	3	3
16	6	5	5	7	7	6	6	5	6	6
17	6	7	7	7	7	7	7	6	7	7
18	5	4	4	7	5	6	6	7	5	7
19	3	4	6	7	5	3	5	3	6	4

FIGURE 1.8
Edit view

Missing Data

Missing data is common in research endeavors and can prove to be problematic for some R functions. At the very least, the researcher using R must be cognizant of the presence of missing data in a file, and the need to properly handle it. Perhaps the most common method for dealing with missing data is simply to delete individuals from the data set who have it. This can be done using the na.omit function in R. Missing data in R should be coded as "NA". The na.omit function removes all cases with the "NA" code in a data set. As an example, if we want a "clean" data set including no observations with missing data, we could use the following command to create it:

```
test.nomiss<-na.omit(test)
```

We do want to point out that such listwise deletion is not often the optimal method for dealing with missing data, and that there exist in R a number of functions designed specifically for this purpose. These are, however, beyond the purview of this book to discuss. In addition, the reader should be aware that different R functions deal with missing data in different ways, so that it is of key importance to learn how a particular function will handle missing observations before using it.

Variable and Data Set Types

Each function in R is designed to handle data of specific types and will not work properly if the data are of another type. Typically (though by no means always) data are either numeric or factor. One shorthand way to think of these is that numeric variables are treated as continuous, while factor variables are treated as categorical. There are greater complexities associated with these types than this simple heuristic would imply, but it serves as a good starting point for our consideration of data types. When we read data into R, numbers will be assumed to be numeric and characters (e.g., letters) will be treated as factors. Generally speaking, this works well for the latter but not always for the former. As an example, if we have a variable indicating the gender of a subject, with males coded as 1 and females as 2, we may want to convert this numeric variable to a factor for use with various statistical analyses. If we leave the data as is, gender will be treated as numeric. This conversion is carried out easily using the as.factor function, as follows:

```
gender.factor<-as.factor(gender)
```

The gender.factor would then be treated as a categorical, or factor variable. We can also convert a variable to numeric if for some reason it is not. For example, if the variable age was read in as a list, rather than numeric, we can use the as.numeric function as follows:

```
age.numeric<- as.numeric(age)
```

A final point to note here is that, as with missing data, different functions have different requirements and conventions when it comes to the types of variables that they expect to see. Therefore, it is important to read the documentation for a particular function in order to determine whether a factor is required or not.

   In addition to the variables being of various types, it is also the case that data objects are also represented as a specific type. Perhaps the most common data type is the dataframe, which corresponds closely to data files readers may be familiar with in SPSS and SAS. In some instances, functions call for data to be in the form of matrices. While to the human user a matrix data format looks quite similar to a dataframe, R sees them as very different objects. If a function calls for the data to be in matrix format, we can use the as.matrix function to convert a dataframe to a matrix:

```
full_data.mat<-as.matrix(full_data).
```

Other data types that may come up in practice are vectors and lists. When using a function, it is important for the analyst to take note of the data type that is required.

Descriptive Statistics and Graphics in R

There are a wide variety of functions available for obtaining descriptive statistics in R. The Hmisc library includes the describe function, which produces several statistics of interest. For this example, we select the ags _ map variable from full _ data and obtain descriptive statistics after we first load the Hmisc library.

```
library(Hmisc)
describe(full_data$ags_map)
vars      n  mean    sd median trimmed  mad min max range   skew kurtosis   se
X1     1 426 17.59 2.67     18   17.84 2.97   7  21    14  -0.87      0.7 0.13
```

We can obtain a histogram for this variable using the following command:

```
hist(full_data$ags_map)
```

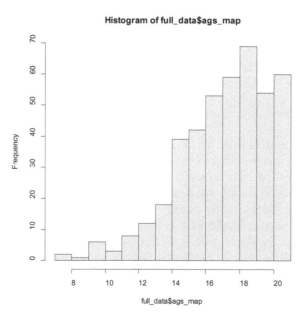

FIGURE 1.9
Histogram of AGS MAP scores

Using the `plot` command, we can obtain a scatterplot for two variables, and by including the `abline` command in conjunction with lm (which we discuss in detail in Chapter 3), we can add a regression line.

```
plot(full_data$ags_map, full_data$ags_mav)
abline(lm(full_data$ags_mav~full_data$ags_map), col="red")
```

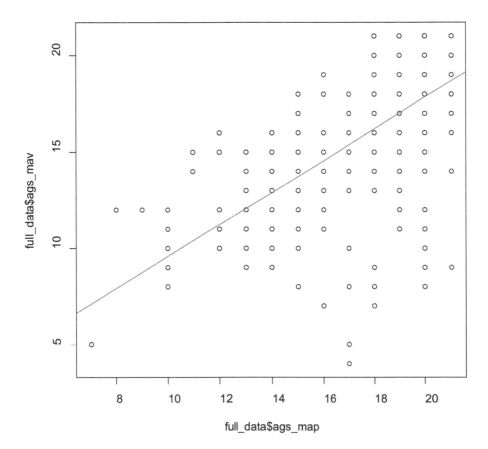

FIGURE 1.10
Scatterplot with regression line for AGS MAV and AGS MAP

Finally, we can use the rcorr function from the Hmdisc library to obtain a correlation matrix, along with the *p*-values associated with testing the null hypothesis that $\rho = 0$ that is associated with each correlation.

```
library(Hmisc)
rcorr(ags_data)
rcorr(ags_data)
```

```
        [,1]    [,2]    [,3]    [,4]
[1,]    1.00    0.65    0.05    0.06
[2,]    0.65    1.00    0.14    0.21
[3,]    0.05    0.14    1.00    0.86
[4,]    0.06    0.21    0.86    1.00
n
        [,1]    [,2]    [,3]    [,4]
[1,]    426     424     426     421
[2,]    424     428     428     423
[3,]    426     428     430     425
[4,]    421     423     425     425
p
        [,1]       [,2]      [,3]      [,4]
[1,]               0.0000    0.3435    0.2577
[2,]    0.0000               0.0050    0.0000
[3,]    0.3435     0.0050              0.0000
[4,]    0.2577     0.0000    0.0000
```

Summary

The purpose of this chapter was to introduce some basic R concepts. This was not intended to be comprehensive, nor is it intended to be the only resource the readers new to R may need for this purpose. In subsequent chapters we will provide further information regarding interactions with R. Indeed, the reader will be provided with all of the information necessary for interacting with the data and conducting the analyses that are demonstrated in this book. In addition, the interested reader will find a plethora of resources for using R, available both online and in textbook form.

In the next chapter, we will turn our attention to the theoretical underpinnings of the regularization methods that are the focus of this book. We will begin by discussing the need for variable selection methods, followed by technical descriptions of each of the estimators that will be the focus of the book. We will conclude the chapter with an examination of inference for regularization methods. These discussions will serve as the foundation for the rest of the text, in which we apply regularization methods in R.

2

Theoretical Underpinnings of Regularization Methods

The purpose of this chapter is to provide a brief introduction to the theoretical foundation of regularization methods such as the lasso, ridge, and elastic net. Our goal is not to give a thorough in-depth discussion of the mathematics of these methods, but rather we want to provide you with sufficient information to understand how these methods work. With this information, you will hopefully be more comfortable when applying these techniques and interpreting work in which they were used. We begin with a review of situations in which regularization and variable selection methods can prove useful, followed by brief technical descriptions of each. We conclude the chapter with a discussion of inference for model parameter estimates for regularization methods. It is important at this point to note that regularization methods are also often referred to as penalized estimators. We will use both terms throughout the book.

The Need for Variable Selection Methods

In some research and evaluation contexts, the number of variables that can be measured (p) approaches or even exceeds the number of individuals on whom such measurements can be made (N). For example, the number of participants in a summer horse camp for children identified with an emotional disability might be relatively small, but the program evaluators may have the capability of measuring a large number of cognitive and affective constructs for each of the participants. Such situations result in what is commonly referred to as high-dimensional data. High-dimensional data present researchers and data analysts with a number of problems, particularly in terms of obtaining stable estimates of model parameters and their associated standard errors.

To be more specific regarding the impact of high-dimensional data, we can consider the widely used linear model framework. This model, which underpins such popular methods as analysis of variance and linear regression, will be discussed in more detail in Chapter 3 of this text. However, before delving into the details, let's consider more generally the impact of small N

DOI: 10.1201/9780367809645-2

and/or large p situations. One major consequence of high-dimensional data is on the estimation of standard errors, which are simply measures of variability associated with model parameter estimates. More specifically, high-dimensional data can yield inflated standard errors for the model coefficient estimates (Bühlmann & van de Geer, 2011). A primary consequence of these inflated standard errors is a reduction in statistical power for testing the model parameter estimates, leading the researcher to erroneously conclude that one or more of the independent variables are not related to the outcome of interest, when in fact they are. Furthermore, having a large number of p independent variables relative to N can result in the presence of collinearity, or very strong relationships among the independent variables (Fox, 2016). Collinearity, in turn, can lead to biased parameter estimates and highly inflated standard errors. In addition to problems of standard error inflation and coefficient bias, high dimensionality can also lead to overfitting of the model to the data, meaning that the parameter estimates do not generalize well beyond the sample upon which they are based (Hastie, Tibshirani, & Friedman, 2011). Finally, when p actually exceeds N, it is simply not possible to obtain estimates for the model parameters using standard methods, meaning that the researcher is not able to address the research questions of interest.

There exist a number of strategies for researchers to use in dealing with high-dimensional data. These include variable selection methods (e.g., stepwise regression, best subsets regression) in which a subset of the variables is selected for inclusion in the final model. These variable selection approaches typically focus on identifying a subset of the independent variables that maximizes the variance explained in the dependent variable (best subsets regression), or that are found to have statistically significant coefficients relating them to the dependent variable (stepwise regression). Another family of methods designed for high-dimensional data is based upon the principle that the p independent variables can be combined into a smaller subset of m linear combinations. In turn, these linear combinations are then treated as the independent variables in a regression analysis with the original dependent variable. Such techniques include principal components regression, supervised principal components regression, and partial least squares regression. Prior research has found that, in the presence of high-dimensional data, the subset selection methods (best subsets and stepwise regression) yield estimates with relatively large variances, leading to somewhat inflated standard errors for the coefficients (Hastie et al., 2011). The data-reduction techniques (principal components, supervised principal components, partial least squares) largely mitigate this problem of inflated variance and standard errors. However, they do so by combining the independent variables into a small number of linear combinations, making interpretation of results for individual variables somewhat more difficult, and creating an extra layer of complexity in the model as a whole (Finch, Hernandez Finch, & Moss, 2014). Thus, although both variable selection and data-reduction techniques can prove useful in the context of high-dimensional data, there remain potential complications that render them less than optimal in a number of situations.

An alternative set of estimators involves incorporating a penalty into the estimation of model parameters. Conceptually, we can think of this penalty as a mechanism for encouraging parsimony in the form of the final model. We will describe the technical details of these methods in more depth below. However, before delving into these equations, we can first consider what is happening from a conceptual perspective. Essentially, these regularization methods take a standard estimation technique, such as least squares or maximum likelihood, and include a penalty factor that is designed to simplify the final model. They do so by driving down the values of model parameter estimates (i.e., penalizing them). Thus, estimates that might have been small when using a standard approach become 0 when using a regularized estimator. Why is this helpful? When several parameters in the model are set to 0, there are fewer values that then need to be estimated, meaning that there is a smaller burden on the available data and fewer opportunities for problems such as estimate instability, inflated standard errors, and collinearity.

A key aspect of applying regularized methods to data is determining the optimal degree to which the parameter estimates should be penalized. We will spend a great deal of time in this book discussing how this is done. We will also see that the penalties described in the rest of this chapter can be applied to a wide variety of statistical methods. Thus, one important point to keep in mind is that the techniques discussed here can be generalized to many different methods, which we will see in subsequent chapters. Finally, we should also keep in mind that there exist several different approaches to regularization and that no one of these can be seen as universally optimal. Indeed, for a given research problem it may not be possible to identify an optimal such approach. Quite often, the methods discussed below will yield similar (though rarely identical) parameter estimates. With those issues in mind, let's look more closely at the regularization techniques that we will be working with together for the rest of the book.

The Lasso Estimator

Regularization methods have in common the application of a penalty to standard statistical estimators such as least squares (LS). One such approach is the least absolute shrinkage and selection operator (lasso, L1; Tibshirani, 1996). The fitting criterion for the lasso penalty as applied to the LS estimator for a linear model is written as

$$e^2 = \sum_{i=1}^{N}\left(y_i - \hat{y}_i\right)^2 + \lambda \sum_{j=1}^{p}\left\lfloor \hat{\beta}_j \right\rfloor \tag{2.1}$$

where

    y_i = Observed value of response variable for individual i

    $\hat{y}$ = Model predicted value of response variable for individual i

    $\hat{\beta}_j$ = Model parameter estimate for variable j; e.g., regression coefficient

    λ = Lasso tuning parameter

The first portion of equation (2.1) is simply the standard least squares estimator, which is minimized in order to find the optimal estimates for model
parameters, such as regression coefficients. The lasso estimator includes the
addition of the tuning parameter λ, which is used to control the amount
of shrinkage (i.e., the degree to which the relationship of the independent
variables to the dependent variable is down-weighted or removed from
the model). Larger λ corresponds to a larger penalty on the estimates; i.e. a
greater reduction in the number of independent variables that are likely to
be included in the final model. When $\lambda = 0$ we have the standard LS estimator. Given the goal of minimizing e^2, the parameter estimates ($\hat{\beta}$) will be
reduced in size when the lasso estimator is applied as compared to standard
LS. In addition, some (perhaps most) of the estimates that would take small
values in LS will be set to 0 when the lasso is applied. Of course, the goal of
any estimator is to yield predictions ($\hat{y}$) based upon the parameter estimates
that are as accurate as possible, meaning that the parameter estimates cannot
all be minimized or set to 0. In other words, the goal of the lasso estimator
is to eliminate from the model those independent variables that contribute
relatively little to the explanation of the dependent variable, by setting their
$\hat{\beta}$ values to 0, while at the same time retaining independent variables that are
important in explaining y.

    The key aspect of using the lasso is the determination of the optimal
λ value. The most common approach to finding the appropriate tuning
parameter value is through the use of cross-validation. With standard cross-
validation, the researcher divides the full sample into k subsamples using
random selection. One of these subsamples is then designated as the training
set, and the others are known as the test sets. The lasso is applied to the training set for a variety of λ values, and the resulting $\hat{\beta}$ estimates are applied
to each of the test samples in order to obtain predicted values of y_i for each
individual. The mean square error for test set k with tuning parameter value
λ ($MSE_{k\lambda}$) is then calculated for each of the test samples as

$$MSE_{k\lambda} = \frac{\sum_{i=1}^{N}\left(y_{ik} - \hat{y}_{ik\lambda}\right)^2}{N_k} \tag{2.2}$$

where

    y_{ik} = Dependent variable value for subject i in test set k

    $\hat{y}_{ik\lambda}$ = Model predicted dependent variable value for subject i in test set k
          using λ

    N_k = Sample size for test set k

The $MSE_{k\lambda}$ values are then averaged across the K test samples for each value of λ, and the optimal value of λ is the one yielding the lowest mean $MSE_{k\lambda}$.

If the sample is too small to be divided into training and cross-validation samples, a variation called leave-one-out or jackknife cross-validation can be used instead. With this method, the lasso model is fit to the data leaving out one individual and then applying the cross-validation method described earlier to compare that individual's actual and predicted values of y. This individual is then placed back into the sample, another individual is removed, the lasso model fit to the data, and model parameters applied to the data of the newly removed individual in order to obtain a cross-validation estimate of the value in equation (2.2). This approach is repeated for each individual in the sample so that the $MSE_{k\lambda}$ is calculated involving all members of the sample. Regardless of which method of cross-validation is used, the optimal value of λ is the one that corresponds to the smallest MSE.

The Ridge Estimator

An alternative to the lasso that uses a very similar penalty function is the ridge estimator. As with the lasso, the ridge penalty is based upon the LS estimator. However, rather than using the sum of the absolute values of the coefficients for the penalty term as in equation (2.1), the sum of the squared coefficients (rather than the absolute values) is used instead. The ridge algorithm seeks to minimize the following fitting function:

$$e_i^2 = \sum_{i=1}^{N}\left(y_i - \hat{y}_i\right)^2 + \lambda\sum_{j=1}^{p}\beta_j^2 \tag{2.3}$$

Although the penalty used in ridge regression looks very similar to that used with the lasso, in fact, they have one primary difference, which is that the lasso serves as both a variable selection procedure and a parameter shrinkage method (Tibshirani, 1996). In practice, this means that the lasso will yield a model in which multiple independent variables may be effectively excluded due to having coefficients of 0. On the other hand, the ridge estimator will drive the coefficients of some variables down to near 0, but will typically not make these coefficients 0 exactly. The result is that ridge models will tend to be less parsimonious than those produced by the lasso (Hastie, Tibshirani, & Wainwright, 2015). Nonetheless, as with the lasso, the ridge estimator seeks to minimize the number of independent variables that effectively contribute to the prediction of the dependent variable, and thereby reduce problems associated with high-dimensional data, such as unstable parameter estimates and collinearity. Determination of the optimal tuning parameter value is done using cross-validation as with the lasso.

The Elastic Net Estimator

The ridge and lasso estimators can be combined in the form of the elastic net (Zou & Hastie, 2005). The elastic net fitting function uses both the absolute squared values and squared coefficients, as well as a second tuning parameter, α:

$$e_i^2 = \sum_{i=1}^{N}\left(y_i - \hat{y}_i\right)^2 + \lambda\sum_{j=1}^{p}\left(\alpha\lfloor\beta_j\rfloor+(1-\alpha)\beta_j^2\right) \tag{2.4}$$

When $\alpha = 0$, equation (2.4) simplifies to the ridge regression model, whereas when $\alpha = 1$ it is the lasso model. The values of α and λ can be selected using cross-validation in the same manner as described earlier.

The Adaptive Lasso

One of the issues associated with using the lasso is bias in the parameter estimates (Hastie et al., 2011). In particular, they have a tendency to be negatively biased, which is a natural outgrowth of the penalty that is applied during the estimation process that drives the estimates down in value and sets some to 0. One approach for countering this bias, while at the same time regularizing the estimates was proposed by Zou (2006), in the form of the adaptive lasso (alasso). The alasso estimator consists of two steps. First, a set of pilot estimates is obtained using the standard lasso, or a regression estimator. These are then applied to the following fitting function:

$$min\left\{\frac{1}{2}\|y - X\beta\|_2^2 + \lambda\sum_{j=1}^{p}w_j\lfloor\beta_j\rfloor\right\} \tag{2.5}$$

where

$$w_j = \frac{1}{\lfloor\tilde{\beta}_j\rfloor^v}$$

$\tilde{\beta}_j$ = Initial lasso estimate for variable j
λ = Penalty term
y = Dependent variable
X = matrix of independent variables
β = Regression coefficients
v = Fitting factor that is greater than 0

The alasso is designed so that parameters with larger initial estimates are generally shrunken less than those parameters with smaller initial estimates.

The alasso yields more severely penalized coefficients for the variables with small parameters in the population, while at the same time yielding less biased estimates for the parameters more generally (Zou, 2006). Thus, the alasso presents itself as a viable alternative to the standard approach. Cross-validation of the usual type is used to determine the optimal value of λ.

The Group Lasso

In some contexts, we may wish to consider variables as belonging to coherent sets such that if one is included in a model, all should be included. For example, a categorical independent variable with three or more categories is typically dummy coded in the context of regression. This dummy coding involves the selection of one category as the reference, and the coding of the other categories as separate variables with a 1 denoting membership in the group, and 0 if not. Consider a variable for membership in one of three treatment groups, control, treatment 1, and treatment 2. If we want to include this variable in a regression analysis, we would need to recode it into two dummy variables. If we select the control group as the reference category, then we would have one variable called treatment1 where an individual has the value of 1 if they are in treatment 1, and 0 otherwise. Similarly, we would also have a variable called treatment2, with a value of 1 for those in treatment 2, and 0 otherwise. Individuals in the control group will have 0 for both variables. If we use this variable in conjunction with the lasso estimator then we need to treat these treatment variables as a single entity, so that if one is included in the final model the other is as well. In other applications, researchers may wish to treat conceptually related variables as belonging to groups. For instance, an epidemiologist examining antecedents of disease outbreaks could group variables into sets based upon what they measure, so that indicators of socioeconomic status are placed in one group, measures of access to health care services in another, health-related behaviors in a third, and family and genetic factors in a fourth group.

Yuan and Lin (2006) described the grouped lasso, which allows variables to be grouped a priori so that the lasso will treat them as a unit. The penalty is then applied to the group, rather than to each variable individually. In order to use the grouped lasso, we first divide the p independent variables into L groups. The grouped lasso algorithm then minimizes the following criterion:

$$min\left(\left\| y - \beta_0 1 - \sum_{l=1}^{L} X_l \beta_l \right\|_2^2 + \lambda \sum_{l=1}^{L} \sqrt{p_l} \left\| \beta_l \right\|_2 \right) \qquad (2.6)$$

where
    p_l = the number of variables in group l
    X_l = the set of predictors in group l

β_0 = Model intercept
β_l = Coefficient vector for group l
λ = Regularization parameter as in the standard lasso
$\sqrt{p_l}$ accounts for size of group l
$\|\beta_l\|_2$ = Euclidean norm of a vector

The set of coefficients for variables in group l, $\|\beta_l\|_2$, is shrunken to 0 only if all components are 0. In this way, the grouped lasso estimates tend to sparsity at both the group and individual variable levels; i.e., group sets of estimates shrink toward 0 in conjunction with the individual estimates within the variable groups. The grouped lasso can be generalized to include the Ridge estimator. In addition, it is possible for groups to overlap with one another, as described in Zhao, Rocha, and Yu (2009).

Bayesian Regularization

A continuing area of research with regularization methods has focused on the development of inferential methods. As we will discuss in the next section, inference for the lasso, ridge regression, and other shrinkage methods is a complex issue and one that has not been fully resolved. One approach for dealing with inference involves the use of Bayesian estimation, rather than the least squares or maximum likelihood approaches that are commonly used with regularization methods. We do not have the space in this book to delve deeply into the topic of Bayesian estimation more generally, given its depth and scope. However, it is necessary for us to briefly review key principles in the most common method for estimating parameters in the Bayesian context, Markov Chain Monte Carlo (MCMC).

Before outlining MCMC, we must first briefly review the core concept of Bayesian parameter estimation, namely the combining of prior information about the parameter estimates with information from our sample in order to create a posterior distribution for the parameter. The prior represents the researcher's beliefs about the likely distribution of the parameter in the population. We will discuss priors in more detail next, but conceptually they can simply be thought of as what form we expect the parameter distribution to take, including the central tendency, variation, and overall shape. In addition to the prior, we will have information from our data. The data might reinforce our prior beliefs, refute them, or fall somewhere in between. The prior and the data are then combined to create the posterior distribution of the parameter, which serves as our estimate of the population distribution. From this distribution, we are able to obtain a measure of the central point using the mean, median, or mode, the variance associated with the parameter (the variance of the posterior distribution), and a credibility interval within which we have some level of confidence that the middle of the distribution falls. Though not identical to the confidence interval with which we are familiar from frequentist statistics, the credibility interval is similar from an interpretive standpoint. We might select the 2.5th and 97.5th percentiles of the posterior distribution as our credibility interval, and then use this value to ascertain whether some value, such as 0, falls within it. This use of the credibility interval serves as our method of inference for the parameter in the Bayesian context.

A key component of conducting Bayesian analysis is the specification of a prior distribution for each of the model parameters. These priors can be either one of two types. Informative priors are typically drawn from prior research and will be fairly specific in terms of both their mean and variance. For example, a researcher may find a number of studies in which a vocabulary test score has been used to predict reading achievement. Perhaps across these studies, the regression coefficient is consistently around 0.5. The researcher may then set the prior for this coefficient as the normal distribution with a mean of 0.5 and a variance of 0.1. By doing so, they are stating up front that the coefficient linking these two variables in their own study is likely to be near this value. Of course, such may not be the case, and because the data is also used to obtain the posterior distribution, the prior plays only a partial role in its determination. In contrast to informative priors, noninformative (sometimes referred to as diffuse) priors are not based on prior research. Rather, noninformative priors are deliberately selected so as to constrain the posterior distribution for the parameter as little as possible, in light of the fact that little or no useful information is available for setting the prior distribution. As an example, if there is not sufficient evidence in the literature for the researcher to know what the distribution of the regression coefficient is likely to be, they may set the prior to be normal with a mean of 0 and a large variance of perhaps 1000, or even more. By using such a large variance for the prior distribution, the researcher is acknowledging the lack of credible information regarding what the posterior distribution might be, thereby leaving the posterior distribution largely unaffected by the prior, and relying primarily on the observed data to obtain the parameter estimate.

Bayesian estimation of the model parameters is typically done using MCMC, which is an iterative process in which the prior distribution is combined with information from the actual sample in order to obtain an estimate of the posterior distributions for each of the model parameters (e.g., regression coefficients, random effect variances). From this posterior distribution, parameter values are simulated a large number of times. After each such sample is drawn, the posterior is updated by including the new value. This iterative sampling and updating process is repeated a very large number of times (e.g., 10,000 or more) until there is evidence of convergence regarding the posterior distribution; i.e., a value from one sampling draw is very similar to the previous sample draw. The Markov Chain part of MCMC reflects the process of sampling a current value from the posterior distribution, given the previous sampled value, while Monte Carlo reflects the random simulation of these values from the posterior distribution. When the chain of values has converged, we are left with an estimate of the posterior distribution of the parameter of interest (e.g., regression coefficient). At this point, a single model parameter estimate can be obtained by calculating the mean, median, or mode from the posterior distribution.

When using MCMC, the researcher must be aware of some technical aspects of the estimation that need to be assessed to ensure that the analysis has worked properly. The collection of 10,000 (or more) individual parameter estimates form a lengthy time series, which must be examined to ensure that two things are true. First, the parameter estimates must converge, and second, the autocorrelation between observations in the final posterior distribution should be low. Parameter convergence can be assessed through the use of a trace plot, which is simply a graph of the parameter estimates in order from the first iteration to the last. The autocorrelation of estimates is calculated for a variety of iterations, and the researcher will look for the distance between estimates at which the autocorrelation becomes quite low. When it is determined at what point the autocorrelation between estimates is sufficiently low, the estimates are thinned so as to remove those that might be more highly autocorrelated with one another than would be desirable. So, for example, if the autocorrelation is low when the estimates are 10 iterations apart, the time series of 10,000 sample points would be thinned to include only every 10th observation, in order to create the posterior distribution of the parameter. The mean/median/mode of this distribution would then be calculated using only the thinned values, in order to obtain the single parameter estimate value. A final issue in this regard is what is known as the burn-in period. Thinking back to the issue of distributional convergence, the researcher will not want to include any values in the posterior distribution for iterations prior to the point at which the time series converged. Thus, iterations prior to this convergence are referred to as having occurred during the burn-in period and are not used in the calculation of posterior means/medians/modes.

Now that we have a basic sense of how parameter estimates are obtained in the context of Bayesian estimation, we can turn our attention to the specifics of the Bayesian lasso. Park and Casella (2008) suggested that the judicious use of prior distributions for regression coefficients will yield shrunken estimators that are comparable to those obtained using the lasso and other methods. And indeed, they demonstrated that this is indeed the case. For example, they recommended the following priors for the response variable and regression coefficients:

$$y \mid \beta, \lambda, \sigma \sim N\left(X\beta, \sigma^2 I_{NxN}\right)$$

$$\beta \mid \lambda, \sigma \sim \prod_{j=1}^{p} \frac{\lambda}{2\sigma} e^{-\frac{\lambda}{\sigma}|\beta_j|}$$

where
  y = Dependent variable
  λ = Tuning parameter
  X = Independent variables
  β = Regression coefficients
  σ^2 = Variance of the error term

The prior distribution for the regression coefficients is typically selected to be the Laplace, which is centered tightly around 0. Thus, the data would need to contain sufficient counter-vailing information regarding these coefficients in order to yield a posterior distribution centered away from 0. Figure 2.1 displays the Normal (0,1) and Laplace distributions with a mean of 0.

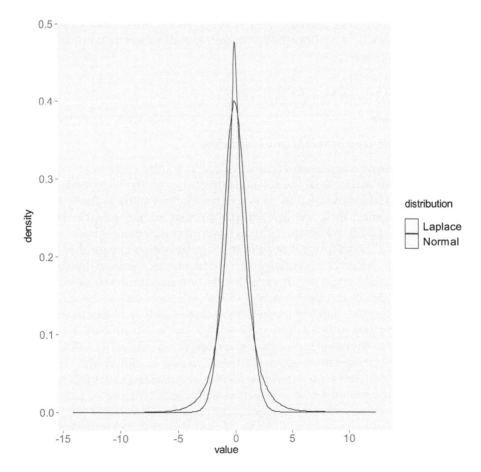

FIGURE 2.1
Normal and Laplace distributions

The more peaked distribution is the Laplace, and we can see that though both are centered on 0, the Laplace is somewhat narrower and more peaked than is the normal.

The point estimate for each model parameter corresponds to the center point of its posterior distribution. The researcher may choose the median, mean, or mode of the posterior for this purpose. Similarly, the standard error of the estimate can be calculated as the standard deviation of the posterior. And, as discussed previously, inference about the parameter can be made using the 95% credibility interval. If 0 falls within the credibility interval for a specific coefficient, then we would conclude that there is no relationship between the variable associated with that coefficient and the dependent variable. Finally, determining whether the Bayesian estimator has converged properly for all model parameters is not a trivial matter. It requires the researcher to examine a variety of statistics and plots in order to ensure that the posterior distributions for the various parameters have all been correctly characterized. We will describe these approaches in more detail in the next chapter when we demonstrate the use of the regularization methods with the R software package.

Inference for Regularization Methods

Researchers using regression techniques are typically interested not only in obtaining an estimate of the relationships between the independent and dependent variables but also in ascertaining whether there is likely to be a relationship among these variables in the population; i.e., whether there is a statistically significant relationship between the independent and dependent variables. The adaptive nature of the regularization approaches makes the question of inference potentially difficult to answer, because the methods are simultaneously engaging in variable selection and parameter estimation (Hastie et al., 2015). In other words, variable selection and statistical inference are intertwined, making the determination of statistical significance for retained coefficients difficult. Researchers working on this problem have suggested using the Bayesian approach that we described earlier (Park & Casella, 2008), or a bootstrap-based approach (Meinhausen & Bühlmann, 2010) in order to conduct statistical inference for the regularization methods. Both approaches incorporate variable selection with model inference, so that the issue of statistical significance remains intertwined with variable selection.

Work has also been done in the area of post-selection inference for regularization methods. Perhaps the most promising of these approaches is the covariance test (Lockhart, Taylor, Tibshirani, & Tibshirani, 2014). This test is conducted after the optimal value(s) of tuning parameters (λ and α) have been selected, and the final set of independent variables to be included in the model has been determined. In order to test for the significance of the coefficient associated with the independent variable x_k, the algorithm first identifies the value of λ for which x_k entered the model, which is denoted as λ_k. The model parameter estimates at this step are then denoted as $\hat{\beta}(\lambda_k)$. Next, the independent variables that were included in the model prior to the entrance of x_p for λ_{k-1} are identified and called A_{k-1}. The algorithm then refits the regularized regression model using only the A_{k-1} set of independent variables (i.e., excluding x_k) but using the value of λ_k as the smoothing parameter. This model yields the parameter estimates $\hat{\beta}_{A_{k-1}}(\lambda_{k+1})$. The covariance test is then calculated as

$$T_k = \frac{1}{\sigma^2}\left(\langle y, X\hat{\beta}(\lambda_k)\rangle - \langle y, X\hat{\beta}_{A_{k-1}}(\lambda_k)\rangle\right) \tag{2.7}$$

where

$\langle y, X\hat{\beta}(\lambda_k)\rangle =$ Covariance between actual y and model predicted y including variable x_k

$\langle y, X\hat{\beta}_{A_{k-1}}(\lambda_k)\rangle =$ Covariance between actual y and model predicted y excluding variable x_k

$\hat{\sigma}^2 = \frac{1}{N-p}RSS_p$

$RSS_p =$ Residual sum of squares for the solution with p predictors

T_k is distributed as an F statistic with 2 and N-p degrees of freedom. We can see that conceptually the covariance test compares the additional amount of variance accounted for by the model when variable x_k is included in the model versus when it is excluded. A statistically significant result for T_k would lead to rejection of the null hypothesis that x_k does not contribute to explaining the dependent variable y; i.e., that x_k accounts for a statistically meaningful amount of variance in the data.

Summary

The goal of this chapter was to introduce the theoretical underpinnings of the various regularization techniques that we will be applying in subsequent chapters. As we will see, these estimators can be incorporated into a wide array of statistical tools, including linear models, generalized linear models, multivariate models, dimension reduction, and multilevel statistical methods, among others. Although they were developed for use in the context of regression, they have since been applied to many other types of problems.

We saw that the basic family of regularization methods, including the ridge, lasso, and elastic net estimators, all have a common framework based upon penalizing model complexity. Models with more non-zero coefficients will tend to fare worse in the selection process than do models with fewer. At the same time, however, these regularization techniques do reward accurate prediction of the response variable(s), meaning that models with too few non-zero coefficients will also fare poorly. This balance of model parsimony and prediction accuracy lies at the heart of all regularization methods.

In this chapter, we also examined common extensions of the lasso, including an approach that allows variables to be grouped in coherent ways, and an adaptation that is designed to reduce estimation bias commonly associated with regularization methods. Finally, we concluded the chapter by describing a Bayesian alternative to the standard estimation approach, which allows for the incorporation of prior information about the parameters, as well as relatively easier model inference. We then discussed more directly the difficulties of model inference with regularized estimation, and a variety of approaches for dealing with these issues. We are now ready to move forward in the next chapter with a demonstration of how these methods can be applied to a standard linear model.

3

Regularization Methods for Linear Models

In this chapter, we will introduce the use of regularization methods for linear models. Our focus will be on the application of these penalized regression models using functions available in the R software package. We will begin this discussion by reviewing the basics of linear regression, and seeing how it can be used in R. We will then look at variable selection methods for regression, including stepwise and best subsets regression. Next, we will turn our attention to the lasso, ridge, and elastic net models and their implementation in R. The extensions to these approaches that we described in Chapter 2, including the adaptive lasso, the grouped lasso, and the Bayesian lasso estimators, and their implementation in R will then be our focus. We will conclude the chapter with a discussion of how to compare the fit of the models with one another, leading to the selection of the optimal approach for a given sample, and a more general discussion of when you might use these methods in practice. By the conclusion of this chapter, you should feel confident in being able to fit the regularized regression estimators in the context of linear models.

Linear Regression

Linear regression is an extremely popular and widely used tool by researchers in fields across many, many disciplines. It allows for an examination of linear relationships between a dependent variable and one or more independent variables. The standard linear regression model can be written as

$$y_i = \beta_0 + \beta_1 x_{1i} + \beta_2 x_{2i} + \cdots + \beta_j x_{ji} \qquad (3.1)$$

where
  y_i = Dependent variable value for subject i
  x_{ji} = Independent variable j value for subject i
  β_0 = Intercept
  β_j = Coefficient for independent variable j

The linear model in equation (3.1) characterizes the relationship between each independent variable, x, with the dependent variable, y, using the

DOI: 10.1201/9780367809645-3

coefficients, β. In order to obtain the estimates for these coefficients ($\hat{\beta}$), the familiar least squares (LS) estimator is typically used. LS identifies $\hat{\beta}$ values that minimize the squared residuals of the regression model in (1), as expressed in equation (3.2).

$$e^2 = \sum_{i=1}^{N}(y_i - \hat{y}_i)^2 \qquad (3.2)$$

where

N = Total sample size

$\hat{y}_i = \hat{\beta}_0 - \hat{\beta}_1 x_{1i} + \hat{\beta}_2 x_{2i} + \cdots + \hat{\beta}_j x_{ji}$

$\hat{\beta}_0$ = Sample estimate of model intercept

$\hat{\beta}_j$ = Sample estimate of coefficient for independent variable j

Put another way, LS seeks to find the values of $\hat{\beta}_0$ and $\hat{\beta}_j$ that minimize the squared difference between the actual dependent variable values and the values that the model predicts. The LS estimator works under three assumptions about the e_i values: (1) e_i follows a normal distribution, (2) the variance of e_i is homogeneous across values of y, and (3) e_i values are independent across observations in the data. When these assumptions are met the LS estimator yields unbiased and efficient estimates of the model parameters.

Fitting Linear Regression Model with R

Now that we have reviewed the basics of linear regression, we can apply it to the example dataset that will be our focus throughout this chapter. These data were collected from 53 college undergraduates and include selected scores from the Wechsler Adult Intelligence Scale (WAIS), and the Kaufman Brief Intelligence Test (KBIT). In addition, each individual was asked to report their mother's level of education, which was reported in one of three categories: high school or less (coded as 1), associate or bachelor's degree (2), and a graduate degree (3). This education variable was also dummy coded into two variables indicating whether the mother had an undergraduate degree or a graduate degree, with the reference category being no college education (discussed in more detail later). The variables included in the data appear in Table 3.1.

In order to demonstrate the fitting of linear regression models in R, we will treat WVB as our dependent variable, with the KBIT scores and mother's education level being the independent variables. We can consider this to be a nominal variable, meaning that we do not want to treat the categories as truly ordered, but rather as labels for the different groups. In other words, we do not want to assume that category 3 is inherently better or more than category 2 or 1.

TABLE 3.1

Variables included in Chapter 3 data

Variable name	Variable label
Mother_Education	Mother's education
WCA	WAIS Cognitive ability standard score
WSM	WAIS Short term memory standard score
WVB	WAIS Verbal standard score
WLR	WAIS Learning standard score
WKN	WAIS Knowledge standard score
KNR	KBIT Number recall standard score
KWO	KBIT Word order standard score
KCT	KBIT Conceptual thinking standard score
KTR	KBIT Triangle standard score
KAT	KBIT Atlantis standard score
KRE	KBIT Rebus standard score
KEV	KBIT Expressive vocabulary
KRD	KBIT Riddles standard score
college	Mother has Associate/Bachelor's degree
post_college	Mother has graduate degree

Therefore, in order to include mother's education level in the analysis, we will need to select one category as a reference against which the other categories will be compared, and then create indicator variables for the remaining categories. In this case, we will select high school or less as the reference, and then create a variable for associate/bachelor's degree (college), which is coded as 1 if the mother has an associate's or bachelor's degree, and 0 otherwise. Similarly, we will create a variable for those having a graduate degree (post_college) with a value of 1 for those whose mother has a graduate degree, and a 0 otherwise. These dummy variables will then be included in the regression model. The coefficient for college is a measure of the mean difference in WVB between students whose mother had an associate/bachelor's degree and those whose mother whose education level was high school or less. The coefficient for post_college reflects the mean difference in WVB between those whose mother had high school or less and those whose mother had a graduate degree.

We can fit the regression model using the lm function, which is part of the MASS library that loads automatically when R is started. The commands for fitting the model appear below. We save the output in an object called chapter3.lm. Within the function call, the dependent variable appears first followed by ~ and then the independent variables separated by +. We then include the name of the dataframe containing the data. The full script for conducting this analysis (and all of the other analyses included in this chapter) appears in the file chapter3.R, which appears on the book website www.routledge.com/9780367408787.

```
chapter3.lm<-lm(WVB~KNR+KWO+KCT+KTR+KAT+KRE+KEV+KRD+college
+post_college, data=chapter3.data)
```

In order to view the output including parameter estimates, standard errors, and hypothesis test results, we apply the summary command to the output object.

```
summary(chapter3.lm)

Call:
lm(formula = WVB ~ KNR + KWO + KCT + KTR + KAT + KRE + KEV +
    KRD + college + post_college, data = chapter3.data)

Residuals:
    Min      1Q  Median      3Q     Max
-4.6369 -1.2952 -0.0528  1.1818  7.2011

Coefficients:
              Estimate Std. Error t value Pr(>|t|)
(Intercept)   32.48072    2.96974  10.937 9.92e-14 ***
KNR            0.10864    0.15859   0.685   0.4972
KWO           -0.15598    0.15311  -1.019   0.3143
KCT            3.63912    0.23937  15.203  < 2e-16 ***
KTR            3.65337    0.15224  23.997  < 2e-16 ***
KAT           -0.03278    0.12835  -0.255   0.7997
KRE           -0.01867    0.14822  -0.126   0.9004
KEV           -0.37121    0.24275  -1.529   0.1339
KRD            0.17492    0.19764   0.885   0.3813
college        0.14734    1.01432   0.145   0.8852
post_college  -2.05639    0.95589  -2.151   0.0374 *
---
Signif. codes:  0 '***' 0.001 '**' 0.01 '*' 0.05 '.' 0.1 ' ' 1

Residual standard error: 2.618 on 41 degrees of freedom
  (1 observation deleted due to missingness)
Multiple R-squared:  0.9814,    Adjusted R-squared:  0.9769
F-statistic: 216.2 on 10 and 41 DF,  p-value: < 2.2e-16
```

The summary command produces a table of summary statistics for the model residuals (difference between the observed and model-predicted dependent variable values), followed by the model coefficients (Estimate), standard errors (Std. Error), test statistics (t value), and *p*-values (Pr(>|t|)). The null hypothesis for each test is that in the population the coefficient is equal to 0; i.e., there is not a relationship between the independent variable and the dependent variable. For this analysis, the coefficients for KCT, KTR, and post_college variables were statistically significantly related to WVB because their *p*-values were less than 0.05. For both KCT and KTR this relationship was found to be positive, meaning that higher scores on the KBIT Conceptual thinking and Triangle test scores were associated with higher WAIS Verbal test scores. The negative coefficient for post_college means that after controlling for the KBIT scores, students whose mother had a graduate degree had lower scores on the WAIS Verbal test than did those whose mother had a high school education or less. Finally, R provides us with information about the overall fit of the model to the data. The Multiple R-squared value of 0.9814 means that 98.14% of the variance in the dependent variable is accounted for by the set of independent variables. When the R^2 value was adjusted for the model complexity (i.e., the number of independent variables), we conclude that the proportion of variance in WVB accounted for by the KBIT scores was 97.69%. Finally, the F-test for the overall model yielded a p-value less than 0.05, meaning that together, the set of KBIT variables was found to be associated with the WAIS Verbal test score.

Assessing Regression Assumptions Using R

As we noted earlier, the use of regression analysis rests upon a set of assumptions about the model error. In particular, we assume that the errors are independently and identically normally distributed. We can assess the assumption of normality for the residuals using a quantile-quantile (QQ) plot. The QQ-plot displays the raw residual values (difference between observed and model-predicted values of the dependent variable) versus the value that they would take if the data were normal. If this plot presents as a straight line, we conclude that the residuals are normally distributed. The QQ-plot for this problem appears in Figure 3.1.

The residuals generally do conform to the normal distribution (represented by the straight line), with the exception of the values at the upper end of the distribution, indicating some positive skewness. Figure 3.2 is the histogram of the residuals, which confirms that indeed there is a slight positive skew to the residuals.

We can also conduct a formal test of the null hypothesis that the data are normally distributed, using the Shapiro-Wilk test.

```
shapiro.test(chapter3.lm$residuals)

        Shapiro-Wilk normality test

data:   chapter3.lm$residuals
W = 0.95128, p-value = 0.03301
```

The p-value of 0.03301 is less than our α of 0.05, leading to the rejection of the null hypothesis and the conclusion that the residuals are likely not normally distributed.

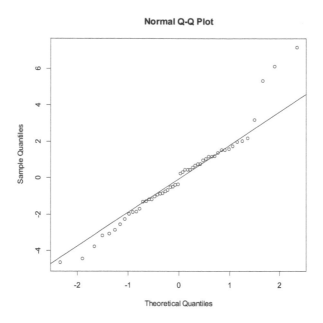

FIGURE 3.1
QQ-plot for residuals of the WAIS verbal test score model

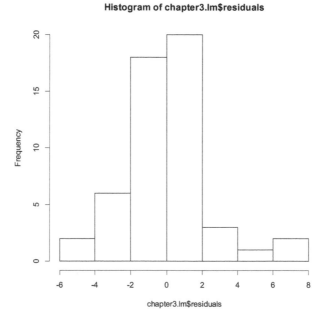

FIGURE 3.2
Histogram of model residuals

A second assumption that we need to check is that the variance of the residuals is homogeneous across the sample. Perhaps the most straightforward way to do this is with a scatterplot of residuals by the model predicted values. The R commands and resulting output appear as follows:

```
plot(chapter3.lm$fitted.values, chapter3.lm$residuals)
```

The assumption of homogeneity of variance appears to be questionable, with greater variability at the upper end of the predicted values.

In addition to the assumptions of normality and homogeneity of the errors, we also need to assess whether collinearity among the independent variables is present. Collinearity refers to the case where there are strong relationships among the independent variables. As we discussed in Chapter 2, the presence of collinearity yields unstable parameter estimates and standard errors, which can in turn lead to incorrect conclusions regarding the relationships

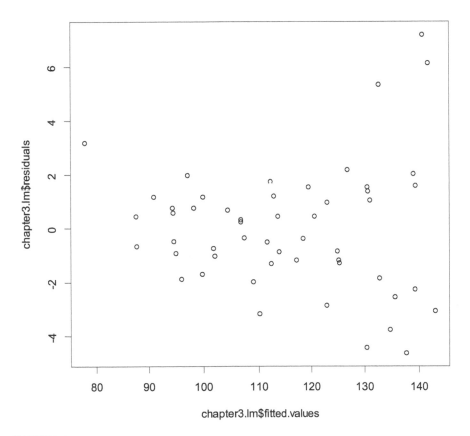

FIGURE 3.3
Scatterplot of residuals by predicted values for assessing homogeneity of errors

among the variables. One of the most straightforward ways to investigate the possibility of collinearity being present is to calculate the variance inflation factor (VIF) for each of the j variables. VIF is calculated as:

$$VIF_j = \frac{1}{1-R_j^2} \qquad (3.3)$$

where

R_j^2 = Proportion of variance for independent variable j accounted for by other independent variables.

Larger values of VIF indicate greater collinearity for the variable. A common cutoff value for VIF is 5, such that variables with a VIF over 5 are considered to exhibit collinearity. We can obtain the VIFs for our model using the `vif` command that is part of the `car` library in R. We simply apply it to the output object from our regression analysis.

```
vif(chapter3.lm)
        KNR         KWO         KCT         KTR         KAT         KRE         KEV         KRD
   1.417355    1.799026    1.901193    1.843510    1.562807    1.448757    1.833667    1.960643
    college post_college
   1.601911    1.668772
```

No variables have VIF of greater than 5, leading us to conclude that there is no collinearity present among our independent variables.

Variable Selection without Regularization

In some cases, we may have a large number of independent variables and/ or a small sample size. As we discussed in chapter 2, this situation, which is referred to as high-dimensional data, can result in problems with model parameter estimation, collinearity, and convergence issues for our estimators. Therefore, it may be necessary to reduce the number of predictors in the model. Of course, one approach for doing so would be to make decisions based upon theory about which variables are most important and thus should be included in the analysis, and which can be removed. However, this may not always be feasible due to uncertainty about which variables are truly important. In such a situation, the researcher may want to make use of a statistical approach for identifying individual independent variables, or combinations of them, that may be most strongly associated with the dependent variable. There exist multiple approaches for doing this, with perhaps the two most popular being stepwise regression and best subsets regression. In the following section of this chapter, we will describe how these approaches for variable selection can be applied using R. It should be noted that there is a broad literature regarding if, when, and how to apply both stepwise and best subsets. We do not have space in this text to delve into those issues, though at the end of the chapter, there is a brief comparison of these techniques with the regularization estimators that are the focus of this text. We refer the interested reader to Fox (2016) and Gelman, Hill, and Vehtari (2021) for excellent summaries of the strengths and weaknesses of variable selection methods.

Stepwise Regression

One of the most common methods for variable selection is stepwise regression. There are two approaches for carrying out stepwise variable selection, forward and backward regression. The forward stepwise algorithm starts with all of the variables outside of the model. Each is then entered into a simple regression model one by one and the hypothesis test results, as well as the R^2 indicating the proportion of variance in y accounted for by this model including x, are saved. The independent variable that accounts for the largest R^2 value, and that has a statistically significant coefficient estimate, is then retained in the model. If none of the independent variables have a statistically significant relationship with y, the algorithm stops and an empty model is returned. Assuming that at least one of the independent variables is retained, then each of the remaining $p-1$ independent variables is then entered in turn into the model including the first retained x, and the R^2 is saved. The x with yielding the greatest increase in R^2 and that has a statistically significant coefficient value is then retained in the model. The forward stepwise algorithm continues until none of the variables outside of the model exhibits a statistically significant relationship with the dependent variable.

Backward stepwise regression begins with all of the independent variables included in the model, and a total R^2 is generated for this full model. Next, each variable is removed in turn and the resulting model R^2 is calculated. The variable that reduces R^2 the least, with no statistically significant change in R^2, is then removed from the model. The algorithm repeats the step with the remaining variables and removes the variable that reduces R^2, the least with a non-significant change in R^2. This process is repeated until there are no variables that can be removed without resulting in a statistically significant reduction in the value of R^2.

Application of Stepwise Regression Using R

Both forward and backward stepwise regression can be employed in R using the step function, which is part of the stats package that is part of the standard R installation. When calling the function, we supply the name of an output object and the stepwise algorithm that we would like to use. Here we will use backward stepwise regression with the chapter3.lm object created earlier, and save the results in an object called chapter3.backward.

```
chapter3.backward<-step(chapter3.lm, direction="backward")
summary(chapter3.backward)

Call:
lm(formula = WVB ~ KCT + KTR + KEV + post_college, data =
chapter3.data)

Residuals:
    Min      1Q  Median      3Q     Max
-4.8773 -1.3118 -0.5109  1.0761  6.9644

Coefficients:
               Estimate Std. Error t value Pr(>|t|)
(Intercept)     32.9479     2.5930  12.706   <2e-16 ***
KCT              3.5858     0.1847  19.419   <2e-16 ***
KTR              3.6535     0.1258  29.049   <2e-16 ***
KEV             -0.2815     0.1766  -1.594   0.1175
post_college    -1.7587     0.7001  -2.512   0.0154 *
---

Signif. codes:  0 '***' 0.001 '**' 0.01 '*' 0.05 '.' 0.1 ' ' 1

Residual standard error: 2.475 on 48 degrees of freedom
Multiple R-squared:  0.9805,    Adjusted R-squared:  0.9789
F-statistic: 604.6 on 4 and 48 DF,  p-value: < 2.2e-16
```

The optimal solution based on backward stepwise selection yielded a model including KCT, KTR, KEV, and post_college. The summary of this model indicates that both KCT and KTR were statistically significantly positively related to WVB, whereas the coefficient for KEV was not statistically significant. Individuals with mothers having a graduate degree had mean WVB scores that were 1.7587 points lower than those whose mothers did not have a graduate degree. Though not shown here, forward selection can also be applied using R by simply replacing backward with forward in the function call. The interested reader will find the code available in the script to be found at the website associated with this book www.routledge.com/9780367408787.

Best Subsets Regression

An alternative approach to variable selection in the context of regression is best subsets regression. With best subsets regression, each possible combination of the independent variables (from single predictor models to the full model containing all predictors) is fit to the data and one or more measures of model fit (e.g., model complexity adjusted R^2, Mallow's C_p, Bayesian information criterion). The model to be retained is the one that maximizes the R^2 estimate after it has been adjusted for the complexity of the model, meaning the number of independent variables. Adjusted R^2, Mallow's C_p, and the Bayesian information criterion (BIC) are calculated as follows:

$$R_A^2 = 1 - \frac{(1 - R^2)(N - 1)}{N - p - 1} \tag{3.4}$$

$$C_p = \frac{\sum_{i=1}^{N}(y_i - \hat{y}_i)^2}{S^2} - N + 2(p + 1) \tag{3.5}$$

$$BIC = -2LL + \ln(N)(p + 1) \tag{3.6}$$

where

R^2 = Proportion of variance in the dependent variable explained by the independent variable
S^2 = Residual variance
LL = Model log-likelihood

Application of Best Subsets Regression Using R

Now that we have been introduced to the concepts underlying best subsets regression, we can apply it to the WAIS verbal test data using the R function regsubsets. We will save the results in an object called chapter3.subsets.models, with the function call requiring the regression equation, the dataframe name, and the maximum number of independent variables that we want to retain. The best subsets algorithm will not search for larger models than that which we request here (8 in this example). We will then request summary information about the model search process.

```
chapter3.subsets.models <- regsubsets(WVB~KNR+KWO+KCT+KTR+KAT+KRE+
KEV+KRD+college+post_college, data = chapter3.data, nvmax = 8)
```

```
summary(chapter3.subsets.models)


Subset selection object
Call: regsubsets.formula(WVB ~ KNR + KWO + KCT + KTR + KAT + KRE +
    KEV + KRD + college + post_college, data = chapter3.data,
    nvmax = 8)
10 Variables   (and intercept)
               Forced in Forced out
KNR                FALSE       FALSE
KWO                FALSE       FALSE
KCT                FALSE       FALSE
KTR                FALSE       FALSE
KAT                FALSE       FALSE
KRE                FALSE       FALSE
KEV                FALSE       FALSE
KRD                FALSE       FALSE
college            FALSE       FALSE
post_college       FALSE       FALSE
1 subsets of each size up to 8
Selection Algorithm: exhaustive

         KNR KWO KCT KTR KAT KRE KEV KRD college post_college
1 ( 1 )  " " " " " " " " " " "*" " " " " " "     " "
2 ( 1 )  " " " " " " " " "*" "*" " " " " " "     " "
3 ( 1 )  " " " " " " " " "*" "*" " " " " " "     "*"
4 ( 1 )  " " " " " " " " "*" "*" " " "*" " "     "*"
5 ( 1 )  " " " " "*" " " "*" "*" " " "*" " "     "*"
6 ( 1 )  " " " " "*" " " "*" "*" " " "*" "*"     "*"
7 ( 1 )  " " "*" "*" " " "*" "*" " " "*" "*"     "*"
8 ( 1 )  "*" "*" "*" "*" "*" " " " " "*" "*"     "*"


best.subsets.results <- summary(chapter3.subsets.models)
data.frame(
   Adj.R2 = which.max(best.subsets.results$adjr2),
   CP = which.min(best.subsets.results$cp),
   BIC = which.min(best.subsets.results$bic)
+)
  Adj.R2 CP BIC
1      4  4   3
```

Based on the adjusted R^2 and Mallow's CP, we would select the 4-variable model as optimal, whereas the Bayesian information criterion (BIC) results suggest that the 3-variable model should be selected. The optimal 4-variable model includes KCT, KTR, KEV, and post_college. The regression results for the model containing only these variables appear as follows:

```
chapter3.lm4<-lm(WVB~KCT+KTR+KEV+post_college,
data=chapter3.data)
```

```
summary(chapter3.lm4)

Call:
lm(formula = WVB ~ KCT + KTR + KEV + post_college, data =
chapter3.data)

Residuals:
    Min      1Q  Median      3Q     Max
-4.8773 -1.3118 -0.5109  1.0761  6.9644

Coefficients:
              Estimate Std. Error t value Pr(>|t|)
(Intercept)    32.9479     2.5930  12.706   <2e-16 ***
KCT             3.5858     0.1847  19.419   <2e-16 ***
KTR             3.6535     0.1258  29.049   <2e-16 ***
KEV            -0.2815     0.1766  -1.594   0.1175
post_college   -1.7587     0.7001  -2.512   0.0154 *
---

Signif. codes:  0 '***' 0.001 '**' 0.01 '*' 0.05 '.' 0.1 ' ' 1
Residual standard error: 2.475 on 48 degrees of freedom
Multiple R-squared:  0.9805,    Adjusted R-squared:  0.9789
F-statistic: 604.6 on 4 and 48 DF,  p-value: < 2.2e-16
```

On the basis of the results of the 4-variable model, we would conclude that KCT and KTR are significantly positively associated with WVB whereas those whose mothers had a graduate degree had WVB scores that were 1.7587 points lower, on average, than those with mothers not having such a degree, after controlling for the other variables. In addition, KEV was not found to be significantly related to WVB.

The results for the 3 variable model appear as follows:

```
chapter3.lm3<-lm(WVB~KCT+KTR+post_college, data=chapter3.data)
summary(chapter3.lm3)

Call:
lm(formula = WVB ~ KCT + KTR + post_college, data =
chapter3.data)

Residuals:
    Min      1Q  Median      3Q     Max
-4.9397 -1.6916 -0.4172  1.2201  6.4394

Coefficients:
              Estimate Std. Error t value Pr(>|t|)
(Intercept)    30.2350     1.9871  15.216   <2e-16 ***
KCT             3.5799     0.1875  19.093   <2e-16 ***
KTR             3.6004     0.1232  29.230   <2e-16 ***
post_college   -1.6866     0.7095  -2.377   0.0214 *
---
```

```
Signif. codes:  0 '***' 0.001 '**' 0.01 '*' 0.05 '.' 0.1 ' ' 1
Residual standard error: 2.513 on 49 degrees of freedom
Multiple R-squared:  0.9795,    Adjusted R-squared:  0.9783
F-statistic: 780.8 on 3 and 49 DF,  p-value: < 2.2e-16
```

With respect to the 3 variable model, KCT and KTR were both found to be significantly positively related to WVB, and individuals with a mother having a graduate degree had a WVB score that was 1.6866 points lower on average than those whose mothers did not have a graduate degree.

One important point to note is that only one of the two dummy coded variables for mother's education was included in the model. This is problematic from a conceptual perspective, as it does not represent the actual construct being measured. Because only the post-graduate variable is included in this model, the college degree and less than college categories are implicitly combined into a single category represented by 0 in the post-graduate variable. Thus, one option for us moving forward would be to force the inclusion of the bachelor's degree indicator variable in the model even though it was not retained by either stepwise or best subsets regression.

Regularized Linear Regression

Now that we have reviewed two approaches for variable selection that do not involve regularization approaches, let's learn how to apply the regularization methods that we described in Chapter 2. There are several libraries available with functions for fitting lasso models using R. We will use the glmnet library because it has a wide array of functions for fitting the lasso, elastic net, and ridge regression models for multiple types of outcome variables. In addition, the glmnet library also provides us with graphical and numeric tools that we can use to thoroughly interrogate our results.

Lasso Regression

Prior to actually fitting the lasso regression model, we need to place the independent variables into a matrix, which will then be passed to the glmnet functions. The following command creates this matrix, which we call chapter3.x. This matrix contains the set of KBIT scores, along with the two dummy coded variables reflecting maternal education level.

```
chapter3.x<-as.matrix(chapter3.data[,7:16])
```

Now we're ready to take an initial look at the lasso regression results, using the following command:

```
chapter3.lasso<-glmnet(chapter3.x, chapter3.data$WVB,
alpha=1, standardize=TRUE, nlambda=100)
```

We will create an output object called chapter3.lasso, using the function glmnet. In the function call, we specify the matrix of independent variables, followed by the dependent variable, the WAIS Verbal test score. We then set the value of α to 1, reflecting that this is lasso regression, as opposed to ridge (where $\alpha = 0$), or elastic net, where α will take some value between 0 and 1. We will request that the predictors be standardized so that comparison of magnitude across variables is possible. Finally, we need to indicate the number of λ values for which we want to fit the models. The glmnet function will fit a separate lasso estimation for each of nlambda values of λs. For each of these models, we will then examine the regression coefficients and R^2 values using both graphical and tabular tools.

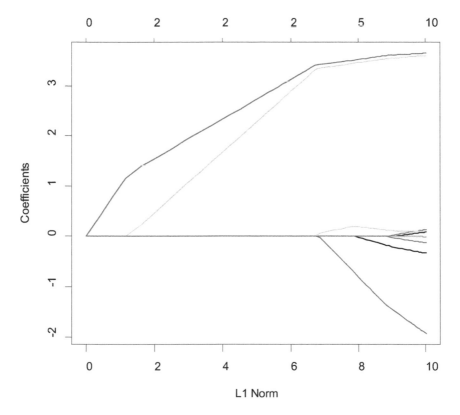

FIGURE 3.4
Lasso regression estimates for independent variables by L1-norm

First, we will examine a plot showing the coefficient values (*y*-axis), by the L1-norm (*x*-axis). Each line in the figure represents the coefficient value for a single variable.

```
plot(chapter3.lasso)
```

We can see that two variables have clear positive coefficients, and an additional variable appears to have a negative regression estimate. The other coefficients included in the model were near 0, regardless of the penalty size.

Figure 3.5 includes the lasso coefficient values (y-axis) by the log of the λ. As we saw in Figure 3.4, there were two variables for which there were clearly non-zero coefficients except for the most severe penalty. In addition, there also appears to be evidence for one variable having a negative coefficient until λ are of a moderate value. None of the other variables had coefficients departing from 0, regardless of the penalty term that was used. These results would suggest that there are probably only three variables in the model that are associated with the response.

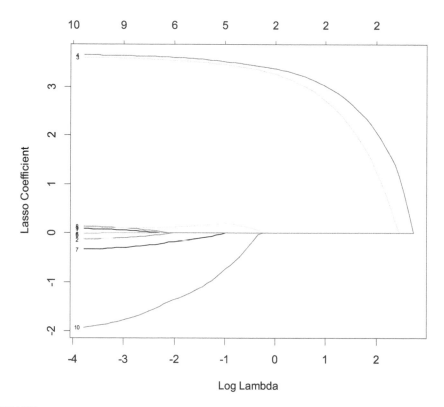

FIGURE 3.5
Lasso coefficient for independent variables by the lasso penalty value

```
plot(chapter3.lasso, xvar="lambda", label=TRUE,
ylab=c("Lasso Coefficient"))
```

The proportion of variance explained in the WAIS Verbal test score (*x*-axis) by the regression coefficients (*y*-axis) appears in Figure 3.6.

```
plot(chapter3.lasso, xvar="dev", label=TRUE, ylab=c("Lasso
Coefficient"), xlab=c("R Squared"))
```

Again, we see that there are clearly two independent variables that appear to be associated with the response, in this case, based on the proportion of explained variance in the response. A third predictor has a somewhat weaker relationship with the dependent variable than do the other two, with none of the other independent variables seeming to be statistically related to the outcome.

On the basis of these results, we would conclude that there is evidence for the applicability of the lasso estimator to our data. Although we have 10 independent variables in the original model, there appear to be only 2 or

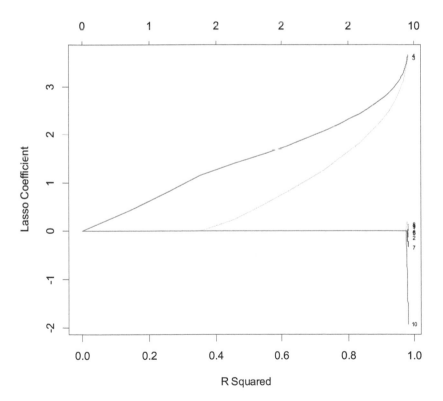

FIGURE 3.6
Lasso coefficient for independent variables by model R^2 value

3 that are actually associated with the dependent variable. Given this finding, we should engage in the next step of fitting our model with the lasso estimator, which involves the selection of the optimal value of λ. This process involves fitting the model with a variety of λ values to a set of k sample cross-validation samples obtained from our original data. For example, if we select k=10, our sample of 53 individuals will be randomly divided into 10 different groups of approximately equal size. For cross-validation group 1, lasso estimates across the range of 100 λ values will be obtained using individuals not in the group, and then predictions of the dependent variable will be obtained for those in group 1 using each estimate model. The same process is repeated for each of the k groups, and then the mean squared error is calculated for each value of λ using the following equation:

$$MSE = \frac{\sum_{i=1}^{n}\left(y_i - \hat{y}_i\right)^2}{n} \tag{3.7}$$

where

y_i = Observed dependent variable value for individual i
$\hat{y}_i$ = Model predicted value for individual i
n = Total sample size

The optimal value of λ corresponds to the smallest *MSE*. The number of k-folds is typically between 5 and 10 but could be as small as n, so that each individual is in their own cross-validation sample.

In order to carry out the cross-validation procedure for selecting the optimal value of λ using R, we will use the following command:

```
chapter3.lasso.cv<-cv.glmnet(chapter3.x, chapter3.data$WVB,
type.measure="mse", nfolds=10)
```

Cross-validation of the lasso estimator is done using the cv.glmnet function. Our output is saved in an object called chapter3.lasso.cv. As with the glmnet function, we specify the matrix of independent variable values, followed by the dependent variable. We then need to specify the measure used to identify the optimal λ value. In addition to the MSE, we could also use the mean absolute error (mae), or the squared error (deviance). Finally, we indicate the number of cross-validation folds as 10. This value can range between 3 and the sample size (53 in our case). There are other options that can be specified when using the cv.glmnet function in specific cases, and which the reader is encouraged to investigate.

We can plot the λ's by the MSE using the following command:

```
plot(chapter3.lasso.cv)
```

The minimum cross-validation MSE corresponds to a Log(λ) of approximately –1.27, which is denoted in the figure by the left-most dashed vertical

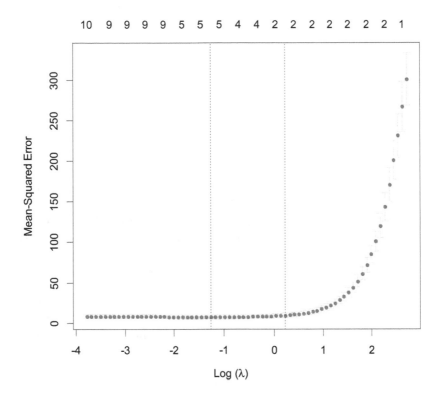

FIGURE 3.7
MSE by Log(λ)

line. The other vertical line corresponds to Log(λ) associated with the MSE that is within 1 standard error of the minimal MSE. We can also obtain the values of λ associated with the minimum and 1SE MSE using the following commands:

```
chapter3.lasso.cv$lambda.min
[1] 0.2807694

chapter3.lasso.cv$lambda.1se
[1] 1.243983
```

Now that we know the optimal value of λ, we can obtain the lasso estimated coefficients associated with it.

```
coef(chapter3.lasso.cv, chapter3.lasso.cv$lambda.min)
11 x 1 sparse Matrix of class "dgCMatrix"
                        1
(Intercept)   32.5171416
KNR                .
```

```
KWO              .
KCT          3.4814915
KTR          3.5425465
KAT              .
KRE              .
KEV         -0.0629726
KRD              .
college      0.1738366
post_college -0.9675080
```

The variables KNR, KWO, KAT, KRE, and KRD all had coefficients that were shrunk to 0. In contrast, KCT, KTR, KEV, college, and post_college all had non-zero coefficients. Notice that the lasso estimates for KCT, KTR, and post_college were somewhat lower than those we obtained using best subsets and stepwise regression, reflecting the negative bias known to be associated with regularization methods.

Now that we have the coefficients for the model based on the optimal λ value, we can investigate the statistical significance of the lasso coefficients, using the techniques described in Chapter 2. We will need to ensure that the selectiveInference library is loaded into R. First, we need to estimate the error standard deviation, which we can do with the estimateSigma command, and save it in an object called lasso.sigma.

```
lasso.sigma<-estimateSigma(chapter3.x, chapter3.data$WVB,
intercept=TRUE, standardize=TRUE)
```

We provide the function with the set of predictors and the dependent variable, indicate that the intercept should be included in the model, and that the variables be standardized. These latter two commands are the default, and they reflect the model that was actually fit to the data. We can see the error standard deviation and degrees of freedom for the model by typing the name of the output object.

```
lasso.sigma
$sigmahat
[1] 2.567819

$df
[1] 6
```

Next, we need to specify the coefficients, which we will save in an object called lasso.beta, and obtain from the following. Note that we explicitly state that the model errors are assumed to be normally distributed family = "gaussian".

```
chapter3.lasso.optimal <- glmnet(chapter3.x, chapter3.
data$WVB, alpha = 1, family = "gaussian",
                lambda = chapter3.lasso.cv$lambda.min)
```

```
lasso.beta = coef(chapter3.lasso.optimal, s=chapter3.lasso.
cv$lambda.min)[-1]
```

We need to exclude the intercept, which is done using [-1] at the end of the `coef` command.

We are now ready to obtain the inference for our model, using the `fixed-LassoInf` command.

```
lasso.inference = fixedLassoInf(chapter3.x, chapter3.
data$WVB,lasso.beta,lambda=chapter3.lasso.cv$lambda.min,
sigma=lasso.sigma$sigmahat)
```

The results are saved in the object called `lasso.inference`. This command requires that we specify the set of independent variables, the dependent variable, the object containing the lasso coefficients, the value of λ for which we want to obtain inference, and the error standard deviation that we obtained earlier. The inference results appears below:

```
lasso.inference

Call:
fixedLassoInf(x = chapter3.x, y = chapter3.data$WVB, beta = lasso.
beta,
    lambda = chapter3.lasso.cv$lambda.min, sigma = lasso.
sigma$sigmahat)

Standard deviation of noise (specified or estimated) sigma = 2.568

Testing results at lambda = 0.281, with alpha = 0.100
```

Var	Coef	Z-score	P-value	LowConfPt	UpConfPt	LowTailArea	UpTailArea
3	3.582	18.174	0.000	3.352	5.822	0.049	0.050
4	3.652	27.567	0.000	3.478	4.756	0.050	0.050
7	-0.279	-1.498	0.075	-1.882	0.078	0.050	0.049
9	0.076	0.077	0.950	-47.309	0.010	0.050	0.050
10	-1.718	-1.919	0.013	-26.974	-1.142	0.050	0.050

```
Note: coefficients shown are partial regression coefficients
```

The output includes coefficient estimates, the z-statistic for testing the null hypothesis that the coefficient is equal to 0 in the population, the p-value associated with that test, the lower and upper bounds of the 90% confidence interval for the coefficient, and the area above the upper and lower tails of this interval. We can change the level of confidence by using `alpha=` subcommand in the function call. Finally, the variables with coefficients penalized to be exactly 0 do not appear in this table. First, we notice that the coefficients that we obtain from the `fixedLassoInf` command are somewhat different from those from `glmnet`. Based on the results presented in this output, we would conclude that variables 3 (KCT), 4 (KTR), and 10 (post_college) were statistically significantly related to the WAIS Verbal score. In contrast, neither variable 4 (KEV) nor variable 9 (college) was found to be associated with the verbal test score, although their coefficients were not exactly 0. We are 90% confident that in the population the KCT coefficient is between 3.352 and 5.822, that the KTR coefficient is between 3.478 and 4.756, and that the post _ college coefficient is between –26.974 and –1.142.

Ridge Regression

In order to fit the ridge regression model, we will use the same set of commands from the glmnet library that we described earlier for the lasso. Given that we discussed the use of those in some detail with respect to the lasso, the following discussion will focus on the results. First, we can

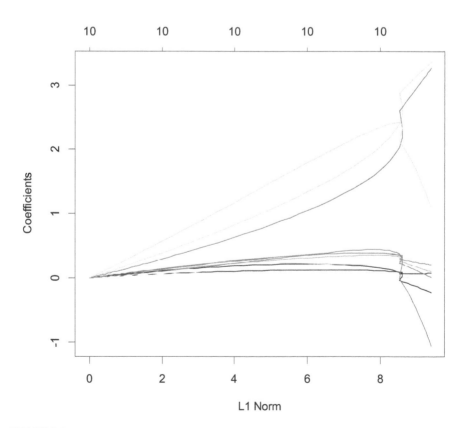

FIGURE 3.8
Ridge regression estimates for independent variables by L1-norm

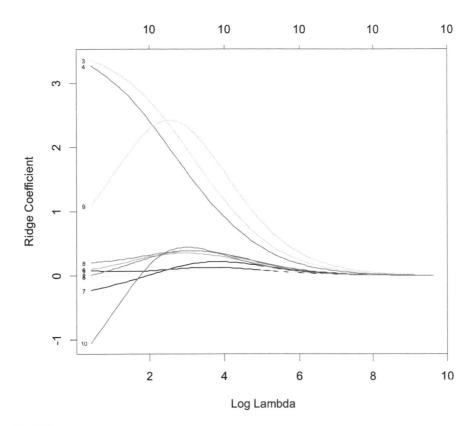

FIGURE 3.9
Ridge coefficient for independent variables by the lasso penalty value

fit the ridge regression model for a set of 100 λ values, with an alpha of 0. By setting alpha to 0 we force the ridge solution, as opposed to using an alpha of 1, which yields the lasso estimator. Following the fitting of the model, we will plot the coefficients by the L1 norm. We see that as the L1 norm value increases, first 3 and then 4 coefficients depart from 0. Thus, it would appear that there are multiple independent variables associated with the WAIS Verbal score. These results diverge somewhat from those of the lasso (Figure 3.4), in that the coefficient values increased more slowly across the L1-norm values in the case of the ridge, and there was also an addition variable with a non-zero coefficient for the ridge than with the lasso.

```
chapter3.ridge<-glmnet(chapter3.x, chapter3.
data$WVB,alpha=0, nlambda=100)

plot(chapter3.ridge)
```

Next, we can ascertain the relationship between the value of the penalty term λ and the coefficient values for the Ridge estimator.

```
plot(chapter3.ridge, xvar="lambda", label=TRUE,
ylab=c("Ridge Coefficient"))
```

The larger the log of the penalty value, the fewer parameter estimates that exceeded 0. Maximally, it appears that for any one value of $\log(\lambda)$, no more than 4 variables had non-zero coefficient values. Again, this is similar, though not identical to what we saw with the lasso.

Finally, in Figure 3.10, we see the Ridge coefficients for the independent variables by the overall model R^2. The proportion of variance explained in the model overall was associated with more non-zero coefficients with the Ridge estimator. However, no more than 4 of the coefficients are non-zero regardless of the tuning parameter.

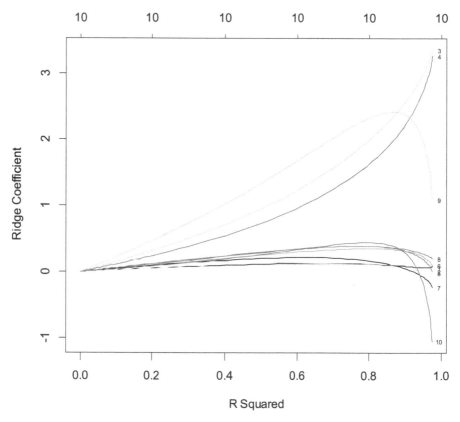

FIGURE 3.10
Ridge coefficient for independent variables by model R^2 value

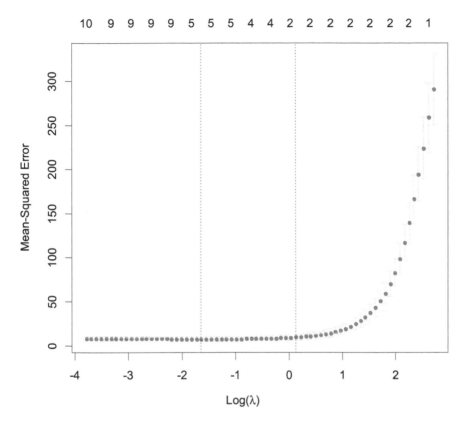

FIGURE 3.11
MSE by Log(λ)

```
plot(chapter3.ridge, xvar="dev", label=TRUE, ylab=c("Ridge
Coefficient"), xlab=c("R Squared"))
```

Now that we have evidence from the previous plots that there are likely to be multiple independent variables with non-zero coefficients, we can identify the optimal λ value for the Ridge estimator. We will start by fitting a cross-validated model with 10 folds and the MSE as the statistic for selecting the optimal model, just as we did for the lasso.

```
chapter3.ridge.cv<-cv.glmnet(chapter3.x, chapter3.data$WVB,
type.measure="mse", nfolds=10)
```

```
plot(chapter3.ridge.cv)
```

The optimal $\log(\lambda)$ is approximately –1.75, and the value that is 1 MSE standard error from the optimal value is approximately 0.1. The exact values of the optimal λ and the 1 SE λ appear as follows:

```
chapter3.ridge.cv$lambda.min
[1] 0.1935235

chapter3.ridge.cv$lambda.1se
[1] 1.133471
```

The coefficients for the Ridge regression estimator at the λ value associated with the minimum cross-validated MSE. These estimates similar in value to those that we obtained using the lasso estimator, although the amount of shrinkage for most of the coefficients is somewhat less for Ridge. For example, both the Ridge KCT and KTR estimates were slightly larger than those from lasso, and both KEV and post_college had slightly more negative Ridge estimates. The one exception to this pattern is for the college variable, in which lasso had a very slightly larger value for the lasso estimator.

```
coef(chapter3.ridge.cv, chapter3.ridge.cv$lambda.min)

11 x 1 sparse Matrix of class "dgCMatrix"
                            1
(Intercept)    32.6498469
KNR                 .
KWO                 .
KCT             3.5126021
KTR             3.5763555
KAT                 .
KRE                 .
KEV            -0.1299300
KRD                 .
college         0.1451011
post_college -1.1998141

chapter3.ridge.optimal <- glmnet(chapter3.x, chapter3.
data$WVB, alpha = 0, family = "gaussian",
                lambda = chapter3.ridge.cv$lambda.min)
```

Now that we have fit the Ridge regression model with the minimum λ value in order to obtain the coefficients, we can obtain the inference results using the same set of commands that we did with the lasso estimator.

```
ridge.beta = coef(chapter3.ridge.optimal, s=chapter3.ridge.cv$lambda.
min)[-1]

ridge.inference = fixedLassoInf(chapter3.x, chapter3.data$WVB,ridge.
beta,chapter3.ridge.cv$lambda.min)

ridge.inference

Call:
fixedLassoInf(x = chapter3.x, y = chapter3.data$WVB, beta = ridge.beta,
    lambda = chapter3.ridge.cv$lambda.min)
```

```
Standard deviation of noise (specified or estimated) sigma = 2.591

Testing results at lambda = 0.194, with alpha = 0.100
```

Var	Coef	Z-score	P-value	LowConfPt	UpConfPt	LowTailArea	UpTailArea
3	3.582	18.014	0.000	3.342	5.703	0.049	0.05
4	3.652	27.325	0.000	3.470	4.697	0.049	0.05
7	-0.279	-1.485	0.080	-1.795	0.090	0.050	0.05
9	0.076	0.077	0.947	-44.736	0.091	0.050	0.05
10	-1.718	-1.902	0.014	-25.609	-1.078	0.050	0.05

```
Note: coefficients shown are partial regression coefficients
```

The results for the ridge regression inference were quite similar to those for the lasso. KCT, KTR, and post_college all had statistically significant relationships with the WAIS Verbal test score. Individuals with higher KCT and KTR scores were associated with higher verbal scores, whereas those whose mothers had a graduate degree had a lower mean verbal test score.

Elastic Net Regression

Given that we have seen how to employ both the lasso and Ridge estimator with R, we can move on to fit the Elastic net using glmnet. A major difference in fitting the net versus both the lasso and Ridge estimators is that we need to find the optimal values of both λ and α, as opposed to only λ. Whereas the glmnet function allows for an automatic search of λ values, using the nlambda command, it does not have an equivalent search function for α. For this reason, we will need to do a bit of extra programming to incorporate the search for the optimal α along with the optimal λ value. This is a relatively easy task that we can carry out using the train function from the caret package, as follows:

```
chapter3.net.model <- train(
  WVB~KNR+KWO+KCT+KTR+KAT+KRE+KEV+KRD+college+post_college,
data = chapter3.data, method = "glmnet",
  trControl = trainControl("cv", number = 10),
  tuneLength = 10
)
```

We will save the output of the analysis in an object named chapter3.net. model. We then explicitly write the regression model, followed by an identification of the data to use. The method command indicates that the glmnet function will be used to fit the model. In traincontrol we indicate that we want to use cross-validation (cv) to identify the optimal tuning parameters, and that a total of 10 cross-validation samples will be used, just as was the case for the lasso and Ridge estimators as described earlier. The `tuneLength` command indicates the number of dimensions for each of the tuning parameters in the grid. The value of 10 here means that there will be 10 levels for α and 10 levels for λ, leading to 100 total combinations in the search grid. The more levels, the more refined will be the search, and the longer the algorithm will take to run.

The first several lines of the output produced by the preceding commands appear as follows:

```
chapter3.net.model

glmnet

53 samples
10 predictors

No pre-processing
Resampling: Cross-Validated (10 fold)
Summary of sample sizes: 48, 47, 46, 47, 48, 48, . . .
Resampling results across tuning parameters:
```

alpha	lambda	RMSE	Rsquared	MAE
0.1	0.01636915	2.717326	0.9840378	2.170064
0.1	0.03781487	2.717326	0.9840378	2.170064
0.1	0.08735725	2.717325	0.9840383	2.170058
0.1	0.20180658	2.710196	0.9841302	2.156956
0.1	0.46619938	2.717542	0.9841653	2.179014
0.1	1.07698107	2.808779	0.9831201	2.287295
0.1	2.48796602	3.220868	0.9796335	2.736828
0.1	5.74752434	4.410410	0.9758102	3.777648
0.1	13.27752702	6.936204	0.9690516	5.929444
0.2	0.01636915	2.713733	0.9841518	2.167938
0.2	0.03781487	2.713733	0.9841518	2.167938
0.2	0.08735725	2.711746	0.9841945	2.164395
0.2	0.20180658	2.699696	0.9844898	2.150176
0.2	0.46619938	2.702887	0.9845553	2.173239
0.2	1.07698107	2.781758	0.9838644	2.253284
0.2	2.48796602	3.191065	0.9826021	2.706947
0.2	5.74752434	4.547501	0.9820033	3.777984
0.2	13.27752702	7.470629	0.9850937	6.305746

We have three options for identifying the optimal combination of the two parameters, which appear in this output, the root mean squared error (RMSE), R^2 (Rsquared), and mean absolute error (MAE). By default the RMSE is employed when the following command is used:

```
chapter3.net.model$bestTune

     alpha     lambda
85       1 0.2018066
```

The 85th (of 100) combination, with $\alpha = 1$ and $\lambda = 0.2018066$ was associated with the smallest RMSE value. If we want to identify the optimal tuning parameter values based on R^2 or MAE, we can use the following commands:

```
chapter3.net.model$results[which.min(chapter3.net.model$results$Rsquared),]

     alpha    lambda      RMSE  Rsquared       MAE   RMSESD RsquaredSD     MAESD
90       1  13.27753  14.85637 0.8691154  12.84135 2.290021 0.09074732  1.376569


chapter3.net.model$results[which.min(chapter3.net.model$results$MAE),]

     alpha    lambda      RMSE  Rsquared       MAE   RMSESD RsquaredSD     MAESD
85       1 0.2018066  2.629893 0.9862182  2.092282 0.7436423 0.01249184 0.4881107
```

Using MAE yielded the same solution as did RMSE, whereas R^2 identified combination 90 as optimal, with $\alpha = 1$ and $\lambda = 13.27753$. One interesting point to note here is that based on both RMSE and MAE yielded tuning parameters that were quite similar to those for the Ridge model, with $\alpha = 1$ and $\lambda = 0.2018066$ for the net as compared to $\alpha = 1$ and $\lambda = 0.1935235$ for the Ridge estimator. Finally, to round out our discussion of these results, we see that the standard deviations of RMSE, Rsquared, and MAE also appear along with the values themselves.

We can now apply the optimal tuning parameters to obtain model parameter estimates using the Elastic net. In the following commands, we first fit the model using the optimal elastic net tuning parameter values, and then print out the resulting coefficients. The results are based on the RMSE, and thus we will label the output as `chapter3.net.optimal.rmse`. We set the value of α and λ using the subcommands alpha = `chapter3.net.model$bestTune[1]` and lambda = `chapter3.net.model$bestTune[2]`.

```
chapter3.net.optimal.rmse <- glmnet(chapter3.x, chapter3.
data$WVB, alpha = chapter3.net.model$bestTune[1],
    family = "gaussian",
                 lambda = chapter3.net.model$bestTune[2])

coef(chapter3.net.optimal.rmse)

11 x 1 sparse Matrix of class "dgCMatrix"
                      s0
(Intercept)   32.6339557
KNR                    .
KWO                    .
KCT            3.5102104
KTR            3.5732184
KAT                    .
KRE                    .
KEV           -0.1237971
KRD                    .
college        0.1450499
post_college  -1.1793823
```

These results reflect non-zero coefficients for KCT, KTR, KEV, college, and post_college. Furthermore, the estimates themselves are quite similar to those for the Ridge estimator, which is to be expected given the close similarity of the tuning parameters, as discussed earlier.

Finally, as was the case for both the lasso and Ridge estimators, we can obtain tests of statistical significance using the commands associated with the `selectiveInference` library.

```
net.sigma<-estimateSigma(chapter3.x, chapter3.data$WVB,
intercept=TRUE, standardize=FALSE)

net.beta.rmse = coef(chapter3.net.optimal.rmse, s=chapter3.net.
model$bestTune[2])[-1]

net.inference.rmse = fixedLassoInf(chapter3.x, chapter3.data$WVB,net.
beta.rmse,lambda=0.03781487, sigma=net.sigma$sigmahat)

net.inference.rmse

Call:
fixedLassoInf(x = chapter3.x, y = chapter3.data$WVB, beta = net.beta.rmse,
    lambda = 0.03781487, sigma = net.sigma$sigmahat)

Standard deviation of noise (specified or estimated) sigma = 2.667

Testing results at lambda = 0.038, with alpha = 0.100
```

Var	Coef	Z-score	P-value	LowConfPt	UpConfPt	LowTailArea	UpTailArea
3	3.582	17.496	0.000	3.328	5.580	0.050	0.050
4	3.652	26.539	0.000	3.461	4.640	0.049	0.050
7	-0.279	-1.443	0.091	-1.710	0.120	0.050	0.049
9	0.076	0.074	0.942	-42.127	0.198	0.050	0.050
10	-1.718	-1.848	0.017	-24.205	-0.958	0.050	0.050

Note: coefficients shown are partial regression coefficients

As was true of the parameter estimates, the inference results for the Elastic net are quite similar to those for Ridge regression, with significant results for KCT, KTR, and post_college. More specifically, if we rely on the Elastic net estimator, we would conclude that higher scores on KCT and KTR are associated with higher WAIS Verbal test scores, and individuals whose mothers had a graduate degree had a lower score on the verbal test than did those whose mother had a high school education, after controlling for the other variables in the model.

Bayesian Lasso Regression

In Chapter 2, we discussed an alternative approach to applying regularization based on the Bayesian estimation paradigm. Recall that this approach allows for parameter inference that is naturally associated with the model estimation process (unlike for standard regularization for which the two are separate). In addition, through the application of priors, we are able to directly control the degree of regularization to apply, as discussed in Chapter 2. The R command to fit the Bayesian lasso is blasso, which is part of the monomvn library. We will save the results in an output object called chapter3.blasso. When fitting the model, we need to pass to the function the matrix of independent variables in the form of chapter3.x, which we used with the lasso above, and the dependent variable chapter3.data$WVB. In this example, we request 20,000 iterations to the MCMC algorithm, and thin the resulting distribution so that only every 10th observation is retained. We will establish the length of the burn-in period when we actually request the results, as follows:

```
chapter3.blasso <- blasso(chapter3.x, chapter3.data$WVB,
T=20000, thin=10)
```

We can take a look at the coefficients using the following command. Note that the burn-in period consists of the first 5,000 observations in the set.

```
summary(chapter3.blasso, burnin=5000)$coef

       mu               b.1               b.2               b.3               b.4
Min.   :20.32   Min.   :-0.45764   Min.   :-0.56240   Min.   :2.742   Min.   :2.966
1st Qu.:29.47   1st Qu.: 0.00000   1st Qu.: 0.00000   1st Qu.:3.415   1st Qu.:3.500
Median :31.11   Median : 0.00000   Median : 0.00000   Median :3.557   Median :3.595
Mean   :31.23   Mean   : 0.00183   Mean   :-0.01189   Mean   :3.556   Mean   :3.595
3rd Qu.:32.87   3rd Qu.: 0.00000   3rd Qu.: 0.00000   3rd Qu.:3.695   3rd Qu.:3.690
Max.   :41.99   Max.   : 0.55470   Max.   : 0.44186   Max.   :4.414   Max.   :4.238

       b.5               b.6               b.7               b.8               b.9
Min.   :-0.489980   Min.   :-0.537112   Min.   :-0.97712   Min.   :-0.6181985   Min.   :-2.9107
1st Qu.: 0.000000   1st Qu.: 0.000000   1st Qu.: 0.00000   1st Qu.: 0.0000000   1st Qu.: 0.0000
Median : 0.000000   Median : 0.000000   Median : 0.00000   Median : 0.0000000   Median : 0.0000
Mean   :-0.009319   Mean   :-0.005652   Mean   :-0.06139   Mean   :-0.0005624   Mean   : 0.1929
3rd Qu.: 0.000000   3rd Qu.: 0.000000   3rd Qu.: 0.00000   3rd Qu.: 0.0000000   3rd Qu.: 0.0000
Max.   : 0.361668   Max.   : 0.428958   Max.   : 0.52023   Max.   : 0.7243008   Max.   : 4.3117

       b.10
Min.   :-4.5640
1st Qu.:-1.6943
Median :-0.5665
Mean   :-0.8777
3rd Qu.: 0.0000
Max.   : 2.0541
```

For the intercept and each regression coefficient, we see the minimum, 1st quartile, median, mean, 3rd quartile, and maximum values from the posterior distributions created by the MCMC algorithm. The mean or median are commonly used as the single point estimate for a parameter in the Bayesian context. Therefore, the estimated relationship between KNR (b.1) and the WAIS Verbal test score is estimated to be 0 whether we use the median or the mean. The same is true for the coefficient of KWO (b.2). On the other hand, the means/medians of the coefficients for KCT (b.3) and KTR (b.4) are 3.557/3.556 and 3.595/3.595, respectively. Indeed, if we rely on the median of the posterior distributions, which is common practice (Kaplan, 2014), then outside of KCT and KTR, only the coefficient for post_college was non-zero, with a value of −0.5665.

Before we can take these coefficient estimates at face value, or investigate inference for them, we need to ensure that our model has converged for each parameter. We can assess convergence using a graphical tool known as a trace plot, which places the parameter estimate on the *y*-axis and the observation number in the chain on the *x*-axis. In addition, we can examine the posterior distribution for each parameter using a histogram. Following are plots of each for the first coefficient (b.1).

```
plot(chapter3.blasso$beta[,1])
hist(chapter3.blasso$beta[,1])
```

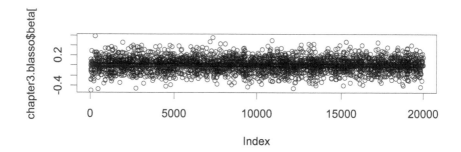

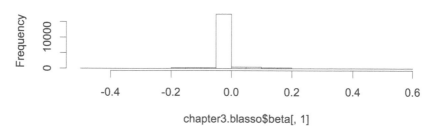

FIGURE 3.12
Trace plot and histogram for coefficient b1

The trace plot appears first, followed by the histogram of the posterior. We are interested in whether the points are centered on a specific point on the y-axis indicating convergence, as opposed to meandering up and down meaning a lack of convergence. These results suggest that indeed the estimates for b.1 have converged, in this case around 0. The histogram shows that the posterior distribution for b.1 is centered at just below 0. In order to ascertain the convergence and posterior distribution of b.3 (KCT), we use the following commands:

```
plot(chapter3.blasso$beta[,3])
hist(chapter3.blasso$beta[,3])
```

Again, the parameter appears to have converged, and the posterior distribution is centered at approximately 3.5. Using the same set of commands for each of the parameters, we see that convergence has been achieved, meaning that we can have confidence in the point estimates and inference (still to be

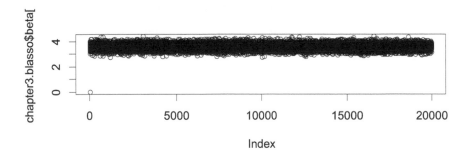

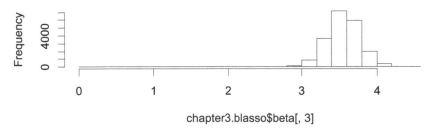

FIGURE 3.13
Trace plot and histogram for coefficient b3

discussed) of the parameters. The same procedure can be used for each of the model parameters.

In addition to assessing the convergence of coefficient estimates, we also may wish to check whether the estimate of the number of non-zero coefficients in the model (*m*) converged as well.

```
plot(chapter3.blasso, burnin=5000, which="m")
```

Again, convergence was achieved. Below are the descriptive statistics for *m*.

```
summary(chapter3.blasso, burnin=5000)$m

  Min. 1st Qu.  Median    Mean 3rd Qu.    Max.
 2.000   3.000   4.000   3.873   5.000  10.000
```

These results indicate that there were likely to be 4 non-zero coefficients in the model, based on the values of the median (4) and mean (3.873).

Histogram of x$m[burnin:x$T]

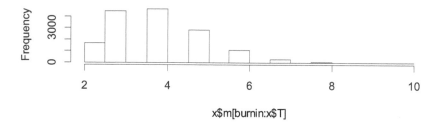

m chain

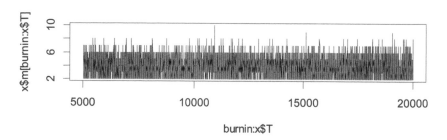

FIGURE 3.14
Trace plot and histogram for *m*

Finally, we can assess the penalty parameter λ in much the same fashion.

```
plot(chapter3.blasso, burnin=5000, which="lambda2")
```

```
summary(chapter3.blasso, burnin=5000)$lambda2
     Min.   1st Qu.     Median      Mean   3rd Qu.       Max.
0.0002461 0.0106528 0.0187983 0.0242486 0.0315415 0.2543454
```

The median of the posterior distribution of λ is fairly small, suggesting that the optimal model does not impose much of a penalty on the coefficients.

Now that we have investigated the parameter estimates themselves, we can turn our attention to the question of inference. In other words, which (if any) of the coefficients can be taken to be significantly different from 0?

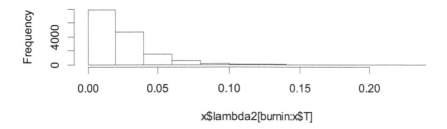

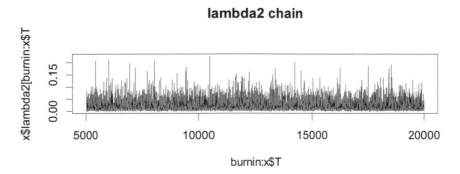

FIGURE 3.15
Trace plot and histogram for coefficient λ

In the context of Bayesian estimation, inference can be examined using the credibility intervals for each posterior distribution. Recall from chapter 2 that the credibility interval corresponds to the portion of the posterior that lies between selected percentiles. For example, we define the 95% credibility intervals as the portion of the posterior distribution that lies between the 2.5th and 97.5th percentiles. Values that lie outside of this interval are considered to be unlikely candidates for the center point of the population distribution of the parameter. Of particular interest is whether 0 lies within the credibility interval. If it doesn't, then in the context of regression coefficients we would conclude that there is a non-zero relationship between the independent and dependent variables.

For our example, we can obtain the credibility intervals using the quantile command.

```
quantile(chapter3.blasso$beta[,1], c(.025, .975))

     2.5%        97.5%
-0.1225360   0.1437973
```

We set the level of confidence by defining the upper and lower percentiles of interest, which at the end of the command. In this case, we define the credibility interval for the first coefficient as lying between the 2.5th and 97.5th percentiles, i.e., the 95% credibility interval. Because 0 is in the interval, we conclude that there is not a statistically significant relationship between KNR and the WAIS Verbal score. We can obtain similar credibility intervals for each of the coefficients using a similar command structure. These values appear in Table 3.2, along with the mean and median of the posterior distributions for each coefficient.

TABLE 3.2

Mean, median, and 95% credibility intervals of the posterior distributions for the lasso regression coefficients

Variable	Mean	Median	95% credibility interval
KNR	0.00183	0	−0.123, 0.144
KWO	−0.01189	0	−0.209, 0.075
KCT	3.557	3.556	3.140, 3.976
KTR	3.595	3.595	3.315, 3.879
KAT	0	0	−0.175, 0.062
KRE	−0.005652	0	−0.168, 0.100
KEV	−0.06139	0	−0.499, 0.022
KRD	−0.0005624	0	−0.176, 0.175
College	0.1929	0	−0.394, 2.039
Post_college	−0.8777	−0.5665	−2.902, 0

On the basis of these credibility intervals, we would conclude that there is a statistically significant relationship between the WAIS Verbal score and both KCT and KTR. This result differs from what we obtained using the lasso in that the coefficient for post_college was not statistically significant.

Bayesian Ridge Regression

We can use the Bayesian approach to fit the Ridge estimator as well. The commands and resulting output are quite similar to what we use with the lasso. Therefore, we will present them here in an abbreviated form. The full R script is available on the book website www.routledge.com/9780367408787. To fit the Bayesian Ridge estimator we will use the `bridge` function that is part of the `monomvn` library.

```
chapter3.bridge <- bridge(chapter3.x, chapter3.data$WVB,
T=20000, thin=10)
```

The coefficients appear as follows:

```
summary(chapter3.bridge, burnin=5000)$coef

        mu              b.1              b.2              b.3             b.4
Min.   :22.25   Min.   :-0.3755752   Min.   :-0.442610   Min.   :2.618   Min.   :3.022
1st Qu.:28.89   1st Qu.: 0.0000000   1st Qu.: 0.000000   1st Qu.:3.437   1st Qu.:3.497
Median :30.38   Median : 0.0000000   Median : 0.000000   Median :3.572   Median :3.585
Mean   :30.47   Mean   :-0.0002342   Mean   :-0.004251   Mean   :3.571   Mean   :3.585
3rd Qu.:31.95   3rd Qu.: 0.0000000   3rd Qu.: 0.000000   3rd Qu.:3.704   3rd Qu.:3.674
Max.   :44.24   Max.   : 0.5017550   Max.   : 0.355599   Max.   :4.346   Max.   :4.185

        b.5              b.6              b.7             b.8              b.9
Min.   :-0.443636   Min.   :-0.491682   Min.   :-0.94765   Min.   :-0.552294   Min.   :-2.5941
1st Qu.: 0.000000   1st Qu.: 0.000000   1st Qu.: 0.00000   1st Qu.: 0.000000   1st Qu.: 0.0000
Median : 0.000000   Median : 0.000000   Median : 0.00000   Median : 0.000000   Median : 0.0000
Mean   :-0.003576   Mean   :-0.002556   Mean   :-0.02379   Mean   :-0.001747   Mean   : 0.1052
3rd Qu.: 0.000000   3rd Qu.: 0.000000   3rd Qu.: 0.00000   3rd Qu.: 0.000000   3rd Qu.: 0.0000
Max.   : 0.323401   Max.   : 0.392416   Max.   : 0.37615   Max.   : 0.477466   Max.   : 3.6431

        b.10
Min.   :-4.912
1st Qu.:-1.173
Median : 0.000
Mean   :-0.553
3rd Qu.: 0.000
Max.   : 1.036
```

We can assess the convergence of parameter estimates using the graphical tools that we discussed earlier with respect to the lasso. As an example, below are the trace plot and histogram for coefficient 3.

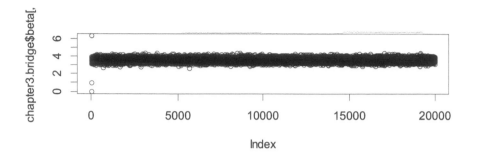

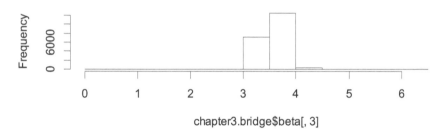

FIGURE 3.16
Trace plot and histogram for coefficient b3

```
plot(chapter3.bridge$beta[,3])
hist(chapter3.bridge$beta[,3])
```

The results for the number of non-zero coefficients based on the Ridge estimator appear as follows:

```
plot(chapter3.bridge, burnin=5000, which="m")

summary(chapter3.bridge, burnin=5000)$m
    Min. 1st Qu.  Median    Mean 3rd Qu.    Max.
   2.000   2.000   3.000   2.755   3.000   7.000
```

As compared to the lasso estimator, which identified approximately 4 non-
zero coefficients, the optimal number was roughly 3 for the Ridge estimator.
    Finally, Table 3.3 includes the mean, median, and 95% credibility intervals
for the Ridge estimator.

Histogram of x$m[burnin:x$T]

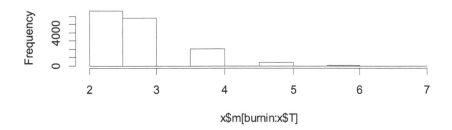

m chain

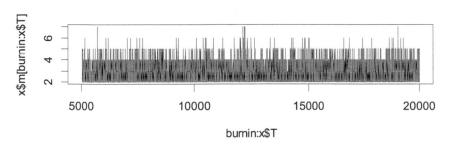

FIGURE 3.17
Trace plot and histogram for coefficient *m*

TABLE 3.3

Mean, median, and 95% credibility intervals of the posterior distributions for Ridge regression coefficients

Variable	Mean	Median	95% credibility interval
KNR	0	0	0, 0
KWO	−0.0002	0	−0.093, 0
KCT	3.571	3.572	3.174, 3.965
KTR	3.585	3.585	3.327, 3.848
KAT	−0.0036	0	−0.073, 0
KRE	−0.0026	0	−0.036, 0
KEV	−0.0238	0	−0.366, 0
KRD	−0.0018	0	0, 0
College	0.1052	0	0, 1.686
Post_college	−0.553	0	−2.707, 0

The primary difference between the Ridge and lasso results is the median for post_college, which is 0 for the former but not the latter. Qualitatively, however, the results for the two approaches were similar, with only KCT and KTR yielding statistically significant results.

Adaptive Lasso Regression

One of the issues that we discussed in Chapter 2 was the bias inherent in the regularized estimators, such as the lasso, Ridge, and elastic net. One approach for dealing with this bias, as described by Zou (2006), is the adaptive lasso (alasso). The alasso takes an initial set of parameter estimates, from the lasso or standard regression, and then applies an additional shrinkage estimate to them. The `glmnet` function provides the user with the ability to fit the alasso model using the `penalty.factor` subcommand. In the following example, we will use the lasso estimates that we discussed earlier as our pilot results, and which we place in the `optimal _ lasso _ coef` object. We use `[-1]` to remove the intercept from the coefficients vector.

```
optimal_lasso_coef <- as.numeric(coef(chapter3.lasso.cv,
chapter3.ridge.cv$lambda.min))[-1]
```

We can now use `cv.glmnet` to fit the lasso again, with the initial set of estimates appearing in the `penalty.factor` subcommand. The terms in this command are multiplied by λ for each variable. We set this factor as 1 divided by the absolute value of the initial lasso coefficient for each variable. In other words, for a given variable λ is multiplied by 1/beta, so that variables with larger initial values will have correspondingly smaller λ 's. Remember from our discussion in Chapter 2 that this was one of the primary features of the alasso, and helps to counter the bias associated with the lasso.

```
alasso_lasso <- cv.glmnet(chapter3.x, chapter3.data$WVB,
                          type.measure = "mse",
                          nfold = 10,
                          alpha = 1,
                          penalty.factor = 1 / abs(optimal_
                          lasso_coef))
```

Our call to the function is quite similar to what we saw earlier in the chapter with regard to the lasso.

First, we can plot the MSE by the log of λ .

```
plot(grp_lasso.fit)
```

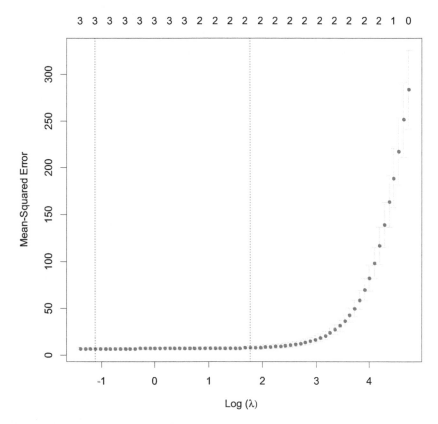

FIGURE 3.18
MSE by log(λ)

The minimum MSE was associated with log(λ) of approximately 1.8. The minimum λ value for the alasso appears as follows:

```
alasso_lasso$lambda.min
[1] 0.3278224
```

In order to obtain the coefficients associated with the optimal model, based on MSE, can be obtained as follows:

```
optimal_alasso_coef <- coef(alasso_lasso, alasso_
lasso$lambda.min)[-1]
```

```
optimal_alasso_coef
```

```
  [1]    0.000000    0.000000    3.564261    3.587762    0.000000
0.000000    0.000000    0.000000    0.000000   -1.407638
```

The alasso coefficients for all variables except KCT, KTR, and post_college were penalized to 0. In addition, the coefficients for those variables that remain in the model were somewhat larger than what we saw with the standard lasso, particularly for post_college. This example clearly demonstrates the impact of using the alasso with two variables (KEV and college) that had non-zero coefficients for the standard lasso being penalized to 0, and the non-zero coefficients being larger for the alasso. The downward bias inherent to regularization methodology was mitigated by the use of the adaptive lasso, as evidenced by the somewhat larger estimates associated with it in this example.

Group Lasso Regression

Recall in Chapter 2 that we discussed situations in which we need to group our variables based on conceptual or practical commonalities among them. For example, we may want to ensure that subscales that measure a common construct be considered together, rather than individually, so that if one is included in the final model they all are. Another case in which the group lasso is useful involves categorical variables that have been dummy coded, such as mother's education in our continuing example. We can consider college and post_college as belonging to a common group, so that if one is included in the model they both are.

We can fit the grouped lasso model using the grpreg function that is part of the grpreg R library. The first step in using the grpreg function is to establish the variable groups. For the WAIS Verbal test example, we want to place the two mother's education variables in a common group, which can be done using the following command to create the group vector:

```
group<-c("KNR", "KWO", "KCT", "KTR", "KAT", "KRE", "KEV",
"KRD", "school", "school")
```

Each variable is assigned to a group where different names refer to different groups. In this example, each of the KBIT variables is in a unique group, and the two mother's education variables are included in a group called "school". *The variable order for the group assignments mirrors their order in the chapter3.x independent variable matrix.* As another example, consider the case where we want to place the first 3 KBIT subscales in a group together, the next 5 KBIT scores in their own group, and then the mother's education variables in a group. In that instance, the group vector would look as follows:

```
group2<-c("kbit1", "kbit1", "kbit1", "kbit2", "kbit2",
"kbit2", "kbit2", "kbit2", "school", "school")
```

We will use the first group vector going forward. After loading the `grpreg` library, we can submit the following command to fit the grouped lasso estimator:

```
grp_lasso.fit<-grpreg(chapter3.x, chapter3.data$WVB, group,
nlambda=100, penalty="grLasso")
```

The results will be saved in the `object grp_lasso.fit`. The command requires the matrix of independent variables, the vector of response variables, the vector identifying the variable groups, the number of lambda values to be assessed (the default is 100), and the penalty to be used. The `grpreg` function allows for a variety of penalty functions, which the interested reader is encouraged to investigate.

A plot of the coefficients reveals a similar pattern to those we have seen before in this chapter, namely that as the penalty decreases in value it appears that a maximum of 3 variables have non-zero coefficients.

```
plot(grp_lasso.fit)
```

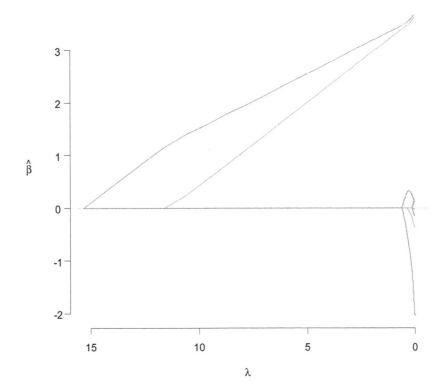

FIGURE 3.19
Coefficient values by λ

We can select the optimal model using one of several statistics, including the AIC, BIC, GCV, AICc, and EBIC. For this example, we will use BIC.

```
select(grp_lasso.fit, "BIC")

$beta
 (Intercept)         KNR         KWO         KCT         KTR         KAT         KRE         KEV
32.38790700  0.00000000  0.00000000  3.46516183  3.52529992  0.00000000  0.00000000 -0.03417134
         KRD     college post_college
 0.00000000  0.30675808  -0.60698303
$lambda
[1]  0.3081441
$df
[1]  5.062206
```

Based on the minimum BIC, the optimal λ was 0.3081441. The coefficients for each variable are also presented through this command. We can see that KCT, KTR, KEV, college, and post_college all had estimates that differed from 0. In addition, note that the estimate for college is larger in the grouped lasso than was true for the standard lasso (0.307 versus 0.17), and the estimate for post_college was closer to 0 than the one obtained using the standard lasso (−0.607 versus −0.968).

Comparison of Modeling Approaches

Now that we have fit a number of estimators to our data, the natural question is, which model provides the best fit? Given that we have used a wide array of approaches (albeit all within the broad framework of regularization methods), the answer to this question is not necessarily easy to obtain. One common approach for assessing model fit is based around the R^2 statistic, which measures the proportion of variability in the dependent variable that is accounted for by the independent variable. We used this statistic early in the chapter when we fit standard regression models to the data. In order to calculate R^2 for a given estimator, we need to first obtain the predicted value of the dependent variable, WAIS Verbal score, based on the estimated coefficients and intercept. We can then calculate the correlation between the predicted and observed dependent variable, and then square this value. In order to adjust for the overestimation of model fit by R^2 when a large number of variables are included in the model, we will also want to examine the adjusted R^2, which includes a penalty for model complexity. For the penalized regression models, only the non-zero parameter values are included in the model complexity penalty parameter, rather than all of the coefficients in the model. There are two very helpful R functions written by Kazuki

TABLE 3.4

R^2 and adjusted R^2 for regression and penalized regression models

Estimator	R^2	adjusted R^2	Non-zero coefficients
Standard regression	0.9813	0.9769	10
Stepwise regression	0.9805	0.9789	3
Best subsets regression	0.9805	0.9789	4
Lasso	0.9792	0.9770	5
Ridge	0.9800	0.9778	5
Elastic net	0.9812	0.9792	5
Adaptive lasso	0.9795	0.9782	3
Adaptive Ridge	0.9760	0.9745	3
Group lasso	0.9796	0.9774	5
Bayesian lasso	0.9753	0.9738	3
Bayesian Ridge	0.9715	0.9704	2

Yoshida to calculate both R^2 and adjusted R^2, and available at https://rpubs.com/kaz_yos/alasso. These functions are also available on the book website: www.routledge.com/9780367408787.

Table 3.4 includes the R^2 and adjusted R^2 values for the various regression and penalized regression models that we have explored in this chapter.

From these results, we see that all of the models yielded extremely good fit to the data, with adjusted R^2 values between 0.97 and 0.98. This means that regardless of the approach we select, approximately 97% to 98% of the variance in the WAIS Verbal test score is explained. When comparing the methods with one another, the elastic net had a very slight advantage over both stepwise and best subsets regression. The Bayesian model had the lowest adjusted R^2 values, though they are very close to those for the other models. On the basis of these results, we would conclude that all of the models yield comparable results, and are able to account for nearly all of the variance in the WAIS Verbal test scores.

Summary

Our focus in this chapter has been the application of regularization and variable selection methods to the problem of multiple linear regression. As we noted at the start, regression is an extremely powerful and commonly used statistical tool that is employed by researchers across many different disciplines. Thus, regularization methods for linear models will have application in many fields. We began our discussion by focusing on the use of stepwise and best subsets regression, which are non-regularized approaches for identifying only the most salient independent variables to include in a regression model. We then moved our attention to the application of the lasso, ridge, and elastic net estimators in R, and saw how to identify whether regularization would be helpful, and if so how to identify the optimal tuning parameter. We then fit these estimators to the data and obtained inference for the resulting parameter estimates. Next, we described a Bayesian approach to the problem of regularization and saw how seamlessly estimation and inference fit together with this approach, when compared to the frequentist methods. Finally, we discussed the adaptive and grouped lasso and saw how relatively easy they are to use in R. In particular, we saw that the negative bias associated with the regularization methods was mitigated by the alasso, making it an attractive alternative. We finished the chapter with a demonstration of how to compare the fits from the many methods examined here in order to select the one that may be optimal for a given problem.

We will close this chapter by briefly discussing the issue of when to use what methods. The short answer to this question is that there is no short answer to this question. Much recent work has been conducted comparing the approaches with one another (see *Statistical Science*, 35(4), November 2020). Taken together, the results of these and other studies suggest that there is not one single approach that will always be optimal in every situation. Rather, we should try several methods and compare the results to one another, as described at the end of this chapter. The simulation studies in Statistical Science suggest that in many situations stepwise and best subsets regression are quite similar. In addition, it is important to note that whereas the regularization methods yield continuous functions relating the independent and dependent variables to one another, stepwise and best subsets don't. In addition, the simulations demonstrated that the variance associated with stepwise and best subsets can be quite large when compared to that of the regularization methods. Thus, rather than thinking in terms of an optimal approach, we should think about the best method for a given research situation.

4

Regularization Methods for Generalized Linear Models

Heretofore, we have focused our attention on models for data in which the outcome variables are continuous in nature. Indeed, we have been even more specific and dealt almost exclusively with models resting on the assumption that the model errors are normally distributed. However, in many applications, the outcome variable of interest is categorical, rather than continuous. For example, a researcher might be interested in predicting whether or not an incoming freshman is likely to graduate from college in 4 years (yes/no), using high school grade point average and admissions test scores as the independent variables. Likewise, consider research conducted by a linguist who has interviewed terminally ill patients and wants to compare the number of times those patients use the word death or dying during the interviews. The number of times that each word appears, when compared to the many thousands of words contained in the interviews is likely to be very small, if not zero for some people. Another way of considering this outcome variable is as the rate of certain target words occurring out of all of the words used by the interviewees. Again, this rate will likely be very low, so that the model errors are unlikely to be normally distributed. Yet another example of categorical outcome variables would occur when a researcher is interested in comparing scores by treatment condition on mathematics performance outcome that are measured on an ordered scale, such as 1, 2, or 3, where higher scores indicate better performance on the mathematics task. For each of these data types, the methods assuming normality of the residuals, which we discussed in Chapter 2, are not applicable. Therefore, we will need to explore alternative model frameworks in order to address questions like those described earlier.

Given that the errors associated with categorical variables that are common in various areas of research will not be normally distributed, we will need to investigate a group of alternative model forms that are collectively referred to as generalized linear models (GLiMs). In the following sections of this chapter, we will focus on three broad types of GLiMs, including those for categorical outcomes (dichotomous, ordinal, and nominal), counts or rates of events that occur very infrequently, and counts or rates of events that occur somewhat more frequently. After describing these models and their standard estimators, we will then turn to the regularized alternatives for them, which are quite similar to those described in Chapters 2 and 3. Once we have

DOI: 10.1201/9780367809645-4

some familiarity with the basic theoretical underpinnings of these GLiMs, we will then describe how they can be fit to data using functions in R.

Logistic Regression for Dichotomous Outcome

Perhaps the most commonly used model for linking a dichotomous outcome variable with one or more independent variables (either continuous or categorical) is logistic regression. The logistic regression model takes the form

$$ln\left(\frac{p(Y=1)}{1-p(Y=1)}\right) = \beta_0 + \beta_1 x \tag{4.1}$$

Here, y is the outcome variable of interest, taking the values 1 or 0, where 1 refers to the outcome of interest; e.g., graduating from college in 4 years. This outcome is linked to an independent variable, x, by the slope (β_1) and intercept (β_0). Indeed, the right side of this equation should look very familiar, as it is identical to the standard linear regression model. However, the left side is quite different from what we see in linear regression, due to the presence of the logistic link function, also known as the logit. Within the parentheses are the odds that the outcome variable will take the value of 1. For the college graduation example, 1 is the value given for graduating in 4 years and 0 is the value for not graduating during that time. By taking the log of these odds, we render the relationship between this outcome and the independent variable (e.g., hours spent studying each week) to be linear. Thus, the logit link for this problem is the natural log of the odds that an individual will graduate from college within 4 years of starting. Interpretation of the slope and intercept in the logistic regression model are the same as interpretation in the linear regression context. A positive value of β_1 would indicate that the larger the value of x, the greater the log odds of the target outcome occurring. The parameter β_0 is the log odds of the target event occurring when the value of x is 0. Logistic regression models can be fit easily in R using the glm function, a part of the MASS library, which is a standard package included with the basic installation of R. In the following section, we will see how to call this function and how to interpret the results we obtain from it.

Fitting Logistic Regression with R

Now that we have a basic understanding of the logistic regression model, let's see how we can fit it using the glm function in R. For this example, we

TABLE 4.1

Independent variables for dichotomous logistic regression analysis

Variable name	Description
eman	Emancipation from family (1=Very dependent, 3=Adequate)
frie	Friends (1=No good friends, 3=two or more good friends)
school	School/employment records (1=Stopped school/work, 3=Moderate/good record
satt	Sexual attitude (1=Inadequate, 3=Adequate)
sbeh	Sexual behavior (1=Inadequate, 3=Adequate)
mood	Mood state (1=Very depressed, 3=Normal)
preo	Preoccupation with weight (1=Completely preoccupied, 3=No preoccupation
body	Body perception (1=Disturbed, 3=Normal)
weight	Body weight (1=Very abnormal, 4=Normal)

will use a dataset containing responses from a sample of 217 people who were diagnosed with an eating disorder, and which can be found at the book website under the name Chapter 4 logistic regression data. The dependent variable (diag2) reflects whether the individual was diagnosed with anorexia (1) or another eating disorder (0). The independent variables for this model appear in Table 4.1.

The basic syntax structure for fitting logistic regression is quite similar to that for standard linear regression. We will save the results of the logistic regression analysis in the object dichotomous.logistic. Within the glm function call, we indicate the dependent variable followed by ~ and then the independent variables separated by +. We then indicate the name of the dataframe to be used, and then the distributional family for the dependent variable. Given that we have two possible outcomes for our dependent variable, the family to be used is the binomial. Once we have fit the model, we then use the summary command to obtain the output, just as with linear regression in Chapter 3.

```
dichotomous.logistic<-
glm(diag2~eman+frie+school+mood+preo+body+satt+sbeh+weight,
data=chapter4.lr.data, family=binomial)

summary(dichotomous.logistic)

Call:
glm(formula = diag2 ~ eman + frie + school + mood + preo + body +
    satt + sbeh + weight, family = binomial, data = chapter4.lr.data)

Deviance Residuals:
    Min       1Q   Median       3Q      Max
-1.6541  -0.9980  -0.7348   1.1703   1.7456
```

```
Coefficients:
              Estimate Std. Error z value Pr(>|z|)
(Intercept)   -0.1126     0.7867   -0.143  0.88619
eman          -0.3047     0.2509   -1.215  0.22452
frie          -0.1574     0.2156   -0.730  0.46541
school         0.2903     0.3112    0.933  0.35101
mood           0.5940     0.2468    2.407  0.01607 *
preo          -0.4083     0.2721   -1.501  0.13342
body           0.2244     0.2524    0.889  0.37381
satt          -0.4427     0.3055   -1.449  0.14733
sbeh           0.4547     0.2866    1.587  0.11256
weight        -0.3506     0.1309   -2.678  0.00741 **
---
Signif. codes:  0 '***' 0.001 '**' 0.01 '*' 0.05 '.' 0.1 ' ' 1

(Dispersion parameter for binomial family taken to be 1)

    Null deviance: 298.38  on 216  degrees of freedom
Residual deviance: 274.81  on 207  degrees of freedom
AIC: 294.81

Number of Fisher Scoring iterations: 4
```

The results of the logistic regression analysis indicate that there was a positive relationship between mood and the log-odds of the likelihood of receiving an anorexia (versus another eating disorder) diagnosis. In addition, there was a negative relationship between weight and the log-odds likelihood of being diagnosed with anorexia. Thus, we would conclude that individuals who exhibited a more normal mood (higher mood scores) were more likely to be diagnosed with anorexia rather than another eating disorders, whereas those who had a more normal weight (higher weight scores) were less likely to be diagnosed with anorexia, versus another eating disorder.

As was true for our OLS regression in Chapter 3, there is a possibility that collinearity could be an issue with the current dataset, particularly given the large number of variables that we've included in the model. We can do this using the vif function from the car library as we did in Chapter 3. It doesn't matter what the dependent variable is, as vif assesses the relationships only among the independent variables. In the following example, we simply fit the linear model using diag2 as the response:

```
dichotomous.linear<-lm(diag2~eman+frie+school+mood+preo+bod
y+satt+sbeh+weight, data=chapter4.lr.data)
vif(dichotomous.linear)
    eman     frie   school     mood     preo     body     satt     sbeh   weight
1.601783 1.261798 2.965914 1.302413 1.902953 1.584977 2.898033 1.684051 1.464115
```

Given that none of our vif values exceeded 5, we can conclude that collinearity is not a problem in this case.

Regularization with Logistic Regression for Dichotomous Outcomes

It is possible to apply the regularization techniques that we discussed in Chapter 2 to the case of logistic regression. For example, the lasso penalty minimizes the following negative log-likelihood function:

$$min\left\{-\frac{1}{N}\mathcal{L}(\beta_0,\beta;y,X)+\lambda\|\beta\|_1\right\} \tag{4.2}$$

where
  N = Sample size
  β_0 = Intercept
  β = Vector of logistic regression coefficients
  y = Vector (length N) of outcomes
  X = Matrix of independent variables
  λ = Penalty parameter.

Similarly, the ridge estimator for logistic regression minimizes the function

$$min\left\{-\frac{1}{N}\mathcal{L}(\beta_0,\beta;y,X)+\lambda\|\beta\|_1^2\right\}. \tag{4.3}$$

The terms in equation (4.2) are as in (4.1). Finally, the elastic net estimator for logistic regression minimizes

$$min\left\{-\frac{1}{N}\mathcal{L}(\beta_0,\beta;y,X)+\lambda\left(\alpha\|\beta\|_1+(1-\alpha)\|\beta\|_1^2\right)\right\}. \tag{4.4}$$

With respect to applying these regularization methods to logistic regression, the same issues and techniques that we discussed in Chapters 2 and 3 are applicable. Specifically, the approaches that we used to determine the optimal values for the regularization parameters, λ and α, are applicable for logistic regression as well. In addition, we will be able to use the same general approach to inference that we demonstrated in Chapter 3. In short, we will see that the principles for applying regularization methods to logistic regression do not differ substantially from OLS regression.

Logistic Regression with the Lasso Penalty

The penalized dichotomous logistic regression model can be fit in R using the `glmnet` function, just as was the case for OLS. First, we need to create a matrix of the predictor variables, which the following command does for the anorexia dataset:

```
chapter4.x<-
as.matrix(chapter4.lr.data[,c("eman","frie","school","mood",
"preo","body","satt","sbeh","weight")])
```

Next, we need to find the optimal value of λ for our sample, using the cv.glmnet function. We can again use k-fold cross validation across a range of 100 λ values. The optimal λ corresponds to the minimization of mean square error (MSE). And, as was the case with OLS regression, we can modify the number of cross-validation folds, number of λ values to investigate, and the statistic to minimize. Following is the R command for fitting a basic lasso estimator to the anorexia data using MSE, 10-fold cross-validation, and a binomial outcome variable.

```
dichotomous.lasso.cv <- cv.glmnet(chapter4.x, chapter4.
lr.data$diag2, alpha = 1, family = "binomial", nfolds=10,
type.measure="mse")
```

We can plot MSE by $ln(\lambda)$ in order to identify the optimal solution.

```
plot(dichotomous.lasso.cv)
```

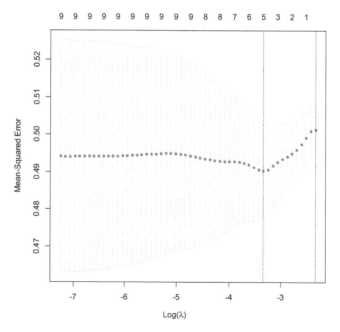

FIGURE 4.1
MSE by $\log(\lambda)$

We can also directly identify the optimal value of λ using the following command:

```
dichotomous.lasso.cv$lambda.min
[1] 0.03539682
```

Thus, based on our 10-fold cross-validation, the minimal value of MSE across the 100 values of λ is 0.03539682.

Now that we have identified the optimal λ, we can use it to obtain the lasso penalized estimates for the logistic regression model. Following is the glmnet call using the optimal λ with the anorexia dataset. We do this much as we did for the linear model, using the glmnet function. Again, we first specify the matrix of predictor variables, followed by the response vector. Given that we are using the lasso penalty, α is set to 1. We indicate that this is a logistic regression model by setting the family to "binomial", and we specify the optimal λ that we obtained from the cross-validation.

```
dichotomous.lasso.optimal <- glmnet(chapter4.x,
chapter4.lr.data$diag2, alpha = 1, family = "binomial",
lambda = dichotomous.lasso.cv$lambda.min)
```

In order to obtain the parameter estimates, we can use the coef command in R.

```
coef(dichotomous.lasso.optimal)
10 x 1 sparse Matrix of class "dgCMatrix"
                        s0
(Intercept)   0.131441536
eman         -0.155775662
frie          .
school        .
mood          0.192878228
preo         -0.001052998
body          .
satt         -0.001856514
sbeh          .
weight       -0.191416728
```

From these results, we can see that the variables frie, school, body, and sbeh were all penalized to 0. In addition, note that the lasso estimates were all smaller than those produced by the standard logistic regression model.

In order to obtain inference for the lasso estimates, we can use functions from the selectiveInference library. Following are the commands for this process, which are essentially the same as those that we used for the linear model in Chapter 3:

```
dichotomous.lasso.sigma<-estimateSigma(chapter4.x, chapter4.
lr.data$diag2, intercept=TRUE, standardize=TRUE)
```

```
dichotomous.lasso.beta = coef(dichotomous.lasso.optimal,
s=dichotomous.lasso.cv$lambda.min)

dichotomous.lasso.inference = fixedLassoInf(chapter4.x, chapter4.
lr.data$diag2,dichotomous.lasso.beta,
 lambda=dichotomous.lasso.cv$lambda.min, family="binomial",
sigma=dichotomous.lasso.sigma$sigmahat)

dichotomous.lasso.inference

Call:
fixedLassoInf(x = chapter4.x, y = chapter4.lr.data$diag2, beta =
dichotomous.lasso.beta,
    lambda = dichotomous.lasso.cv$lambda.min, family = "binomial",
    sigma = dichotomous.lasso.sigma$sigmahat)

Testing results at lambda = 0.035, with alpha = 0.100
```

Var	Coef	Z-score	P-value	LowConfPt	UpConfPt	LowTailArea	UpTailArea
1	-0.163	-0.721	0.019	-Inf	-0.749	0.00	0.05
4	0.605	2.657	0.964	-Inf	-0.272	0.00	0.05
5	-0.262	-1.146	0.992	2.268	Inf	0.05	0.00
7	-0.163	-0.836	0.987	0.847	Inf	0.05	0.00
9	-0.269	-2.230	0.002	-Inf	-0.598	0.00	0.05

```
Note: coefficients shown are full regression coefficients
```

From these results, it appears that coefficients 1 and 9, corresponding to eman (emancipation from family) and weight, were statistically significant. To summarize our results, therefore, we would conclude that individuals who were more adequately emancipated from their families were less likely to be diagnosed with anorexia versus another eating disorder. Similarly, those whose weight was more normal were also less likely to be diagnosed with anorexia.

Logistic Regression with the Ridge Penalty

We can apply the ridge penalty to the logistic regression model using the glmnet function, as was the case with OLS regression. For the anorexia diagnosis data, the ridge estimator is obtained using the following command sequence. Note that all of the steps are included here, from identification of the optimal l value, through inference for the parameters. These steps are identical to those for the lasso, except that we set alpha to 0, rather than 1.

```
dichotomous.ridge.cv <- cv.glmnet(chapter4.x, chapter4.
lr.data$diag2, alpha = 0, family = "binomial", nfolds=10,
    type.measure="mse")
```

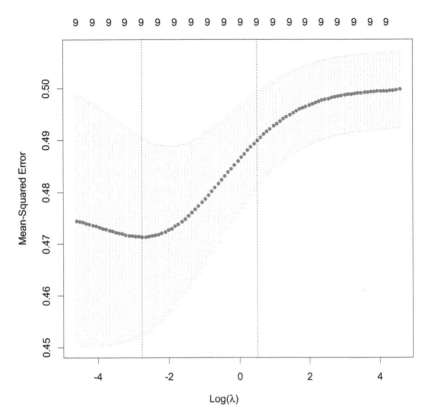

FIGURE 4.2
MSE by $\log(\lambda)$

```
plot(dichotomous.ridge.cv)
dichotomous.ridge.optimal <- glmnet(chapter4.x, chapter4.
lr.data$diag2, alpha = 0, family = "binomial", lambda =
dichotomous.ridge.cv$lambda.min)
coef(dichotomous.ridge.optimal)
10 x 1 sparse Matrix of class "dgCMatrix"
                   s0
(Intercept)   0.1822848
eman         -0.2347508
frie         -0.1237961
school        0.0595543
mood          0.3991488
preo         -0.2489776
body          0.1166265
satt         -0.1953893
sbeh          0.2576972
weight       -0.2371193
```

As was true in the OLS case, the ridge estimator does not penalize all parameters down to 0, but does yield reduced estimates when compared with those from the standard logistic regression model. Following is the R code and resulting output for the inference of these estimates.

```
dichotomous.ridge.beta = coef(dichotomous.ridge.optimal,
s=dichotomous.ridge.cv$lambda.min)

dichotomous.ridge.inference = fixedLassoInf(chapter4.x, chapter4.
lr.data$diag2,dichotomous.ridge.beta,
 lambda=dichotomous.ridge.cv$lambda.min, family="binomial",
sigma=dichotomous.ridge.sigma$sigmahat)

dichotomous.ridge.inference

Call:
fixedLassoInf(x = chapter4.x, y = chapter4.lr.data$diag2, beta =
dichotomous.ridge.beta,
    lambda = dichotomous.ridge.cv$lambda.min, family = "binomial",
    sigma = dichotomous.ridge.sigma$sigmahat)

Testing results at lambda = 0.063, with alpha = 0.100
```

Var	Coef	Z-score	P-value	LowConfPt	UpConfPt	LowTailArea	UpTailArea
1	-0.299	-1.228	0.304	-0.687	0.578	0.049	0.050
2	-0.156	-0.745	0.520	-0.472	0.930	0.049	0.050
3	0.275	0.920	0.759	-4.229	0.570	0.050	0.050
4	0.579	2.449	0.032	0.070	0.967	0.049	0.050
5	-0.396	-1.508	0.229	-0.828	0.485	0.050	0.049
6	0.216	0.884	0.551	-1.339	0.566	0.050	0.049
7	-0.426	-1.450	0.598	-0.789	2.013	0.049	0.000
8	0.445	1.600	0.219	-0.508	0.917	0.049	0.048
9	-0.343	-2.706	0.086	-0.547	0.081	0.049	0.049

```
Note: coefficients shown are full regression coefficients
```

On the basis of these results, we would conclude that only the coefficient for variable 4 (mood) was statistically significant. As we noted earlier, the coefficient was positive, indicating that the more positive the mood, the greater the likelihood of an individual being diagnosed with anorexia versus another eating disorder.

Penalized Logistic Regression with the Bayesian Estimator

As was the case for OLS regression, we can apply a Bayesian approach to fit the logistic regression model with a penalized estimator. The EBglmnet package can be used for either the lasso or elastic net penalties. These approaches

differ in terms of the prior distributions that are used for the model parameters. The default priors for the lasso parameter are

$$\beta_j \sim N\left(0, \sigma_j^2\right) \tag{4.5}$$

$$\sigma_j^2 \sim exp(\lambda) \tag{4.6}$$

In this case, λ is the standard lasso regularization parameter. The default
elastic net priors used in the function take the form

$$\beta_j \sim N\left[0, \left(\lambda_1 + \tilde{\sigma}_j^{-2}\right)^{-2}\right] \tag{4.7}$$

$$\tilde{\sigma}_j^{-2} \sim generalized-gamma\left(\lambda_1, \lambda_2\right). \tag{4.8}$$

The two hyperparameters for the elastic net prior are associated with the
coefficient and the variance.

Finally, the default prior used by EBglmnet, if none is specified by the user,
is the lasso-NEG:

$$\beta_j \sim N\left(0, \sigma_j^2\right) \tag{4.9}$$

$$\sigma_j^2 \sim exp(\lambda) \tag{4.10}$$

$$\lambda \sim gamma(a, b). \tag{4.11}$$

The hyperparameters for λ are $a \geq -1$ and $b > 0$.

Applying the regularized Bayesian estimator to the dichotomous logistic
regression problem is straightforward using EBglmnet. First, we can use the
default set of priors, lasso-NEG.

```
dichotomous.lassoneg.bayes <-
cv.EBglmnet(chapter4.x,chapter4.lr.data$diag2,family="binom
ial", prior=c("lassoNEG"), nfolds=10)

EBLASSO Logistic Model, NEG prior,Epis: FALSE; 10 fold
cross-validation
```

The primary function to be used in fitting the model is cv.EBglmnet. The
function call is quite similar to that of cv.glmnet, requiring that we specify
the matrix of independent variables, followed by the vector for the response.
EBglmnet can be used to fit OLS or dichotomous logistic regression. We specify the type of model through the family subcommand, with "binomial"
indicating logistic regression, and "gaussian" used for OLS regression. We
then indicate the prior distribution to be used, in this case lassoNEG, and the
number of folds for the cross-validation, 10.

In order to obtain the results, we request the fit object from within the broader `cv.EBglmnet` output object, which we have named `dichotomous.lassoneg.bayes`.

```
dichotomous.lassoneg.bayes$fit
      locus1 locus2      beta posterior variance t-value    p-value
[1,]     9      9 -0.2406293       0.0085782 2.598067 0.01002018
```

This object includes columns for the variables and/or interactions that were not penalized to 0, the mean of the posterior distribution for the non-zero coefficients, the variance of the posterior distribution, the *t*-statistic (ratio of posterior mean to the square root of the posterior variance), and the *p*-value associated with this *t*-statistic. These results for this problem indicate that only the coefficient for the 9th variable in the independent set, which corresponds to weight, was not penalized to 0 and was statistically significant. The negative coefficient means that the more normal a person's weight, the less likely that they were diagnosed with anorexia versus other eating disorders, matching the results from the standard penalized estimates.

One of the more useful features of `EBglmnet` is that it can automatically include 2-way interactions among the independent variables. This is done using the `Epis` subcommand, which is `FALSE` by default meaning that no interactions are included. We can include the 2-way interactions for the anorexia data with the `lassoNEG` priors using the following command:

```
dichotomous.lassoneg.interaction.bayes <-
cv.EBglmnet(chapter4.x,chapter4.lr.data$diag2,family="binomial",
Epis=TRUE, prior=c("lassoNEG"), nfolds=10)
```

```
dichotomous.lassoneg.interaction.bayes$fit
      locus1 locus2      beta posterior variance t-value    p-value
[1,]     1      5 -0.1099669       0.003471171 1.866483 0.063329059
[2,]     4      8  0.2047604       0.004554500 3.034070 0.002708602
[3,]     7      9 -0.1194074       0.001652640 2.937257 0.003670288
```

There were three interactions that were not penalized to 0, including between variables 1 and 5, 4 and 8, and 7 and 9. The interactions between 4 and 8, and 7 and 9 were statistically significant at α=0.05. We won't discuss these results further here, but the interested reader is encouraged to delve further into the interpretation of statistically significant interactions, and the complications of doing so with penalized estimators. It is important to note, however, that `EBglmnet` can be a very useful tool for identifying potentially important/interesting 2-way interactions in the data.

We can fit a model using the `lasso` priors, and obtain the results with the following commands:

```
dichotomous.lasso.bayes <- cv.EBglmnet(chapter4.x,chapter4.lr.data
$diag2,family="binomial", prior=c("lasso"), nfolds=10)

EBLASSO Logistic Model, NE prior,Epis:  FALSE ; 10 fold cross-
validation

dichotomous.lasso.bayes$fit
      locus1 locus2       beta posterior variance  t-value   p-value
 [1,]      9      9 -0.035277        0.001237337 1.002877 0.3170423
```

As with the lassoNEG priors, only weight was found not to have 0 parameters based on regularization. However, in this case its relationship with the diagnosis was not statistically significant. In addition, the coefficient was much closer to 0 for the lasso priors than was the case for the lassoNEG.

Finally, we can apply the elastic net estimator with the following command sequence:

```
dichotomous.net.bayes <- cv.EBglmnet(chapter4.x,chapter4.lr.data$d
iag2,family="binomial", prior=c("elastic net"), nfolds=10)

dichotomous.net.bayes$fit
      locus1 locus2       beta posterior variance  t-value    p-value
 [1,]      9      9 -0.2867824        0.01033999 2.820281 0.005244815
```

Again, only weight was not penalized to 0. And as with the lassoNET priors, it was found to be statistically significantly related to the diagnosis, with the negative coefficient meaning that the more normal a person's weight, the lower the likelihood that they would be diagnosed with anorexia versus other eating disorders.

Adaptive Lasso for Dichotomous Logistic Regression

In Chapters 2 and 3, we discussed the use of adaptive regularization methods. Recall that the purpose of this adaptive approach is to deal with estimation bias that is inherent in regularization methods (Zou, 2006). Adaptive regularization attempts to ameliorate this problem by taking an initial set of parameter estimates from an initial analysis (e.g., lasso, ridge, elastic net, standard model) and then applying shrinkage estimates to them. We saw in Chapter 3 that carrying out adaptive regularization for models involving continuous dependent variables is fairly straightforward using the glmnet library. Such is also the case for logistic regression with a dichotomous response. The sequence of R commands necessary to apply the adaptive lasso appears below. Given that these are discussed in detail in Chapter 3, we refer the reader there for a detailed description of how the commands are set up and used.

```
optimal_dichotomous_lasso_coef <- as.numeric(coef(dichotomous.
lasso.cv, dichotomous.lasso.cv$lambda.min))[-1]

dichotomous_alasso_lasso <- cv.glmnet(chapter4.x, chapter4.
lr.data$diag2,
                        type.measure = "mse",
                        nfold = 10,
                        alpha = 1,
                        penalty.factor = 1 / abs(optimal_
                        dichotomous_lasso_coef))

dichotomous_optimal_alasso_coef <- coef(dichotomous_alasso_lasso,
dichotomous_alasso_lasso$lambda.min)[-1]

dichotomous_lasso.sigma<-estimateSigma(chapter4.x, chapter4.
lr.data$diag2, intercept=TRUE, standardize=TRUE)

dichotomous_alasso.inference = fixedLassoInf(chapter4.x, chapter4.
lr.data$diag2,dichotomous_optimal_alasso_coef,
    lambda=dichotomous_alasso_lasso$lambda.min, sigma=dichotomous_
lasso.sigma$sigmahat)

dichotomous_alasso.inference

Call:
fixedLassoInf(x = chapter4.x, y = chapter4.lr.data$diag2, beta =
dichotomous_optimal_alasso_coef,
    lambda = dichotomous_alasso_lasso$lambda.min, sigma =
dichotomous_lasso.sigma$sigmahat)

Standard deviation of noise (specified or estimated) sigma = 0.485

Testing results at lambda = 0.040, with alpha = 0.100
```

Var	Coef	Z-score	P-value	LowConfPt	UpConfPt	LowTailArea	UpTailArea
1	-0.072	-1.493	0.136	-0.150	0.041	0.050	0.049
4	0.117	2.353	0.019	0.027	0.200	0.050	0.049
9	-0.077	-2.834	0.005	-0.123	-0.030	0.048	0.049

After first fitting the initial lasso regression model to obtain starting coefficients, the second round of regularization and associated inference results appear earlier. From these results, we see that three of the variables were not regularized to 0, with variables 4 (mood) and 9 (weight) being statistically significantly related to the log-odds of being diagnosed with anorexia. The positive coefficient for mood means that individuals with a more positive mood were more likely to be diagnosed with anorexia, whereas the negative coefficient for emancipation indicates that those who were more emancipated were less likely to be diagnosed with anorexia.

The adaptive ridge estimator is also an option for use in the context of dichotomous logistic regression. As with the adaptive lasso, we discuss the

use of the adaptive ridge estimator in some detail in Chapter 3, and will use
the following command sequence in the context of logistic regression:

```
optimal_dichotomous_ridge_coef <- as.numeric(coef(dichotomous.
ridge.cv, dichotomous.ridge.cv$lambda.min))[-1]

dichotomous_aridge_ridge <- cv.glmnet(chapter4.x, chapter4.
lr.data$diag2,
                         type.measure = "mse",
                         nfold = 10,
                         alpha = 0,
                         penalty.factor = 1 / abs(optimal_
                         dichotomous_ridge_coef))

dichotomous_optimal_aridge_coef <- coef(dichotomous_aridge_ridge,
dichotomous_aridge_ridge$lambda.min)[-1]

dichotomous_ridge.sigma<-estimateSigma(chapter4.x, chapter4.
lr.data$diag2, intercept=TRUE, standardize=TRUE)

dichotomous_aridge.inference = fixedLassoInf(chapter4.x, chapter4.
lr.data$diag2,dichotomous_optimal_aridge_coef,
    lambda=dichotomous_aridge_ridge$lambda.min, sigma=dichotomous_
ridge.sigma$sigmahat)

dichotomous_aridge.inference

Call:
fixedLassoInf(x = chapter4.x, y = chapter4.lr.data$diag2, beta =
dichotomous_optimal_aridge_coef,
    lambda = dichotomous_aridge_ridge$lambda.min, sigma =
dichotomous_ridge.sigma$sigmahat)

Standard deviation of noise (specified or estimated) sigma = 0.485

Testing results at lambda = 0.425, with alpha = 0.100
```

Var	Coef	Z-score	P-value	LowConfPt	UpConfPt	LowTailArea	UpTailArea
1	-0.069	-1.212	1	5.716	Inf	0	0
2	-0.037	-0.746	1	4.936	Inf	NaN	0
3	0.062	0.888	1	7.020	Inf	0	0
4	0.132	2.415	1	-Inf	-5.468	0	NaN
5	-0.090	-1.464	1	6.130	Inf	0	0
6	0.048	0.838	1	-Inf	-5.718	0	NaN
7	-0.098	-1.416	0	-Inf	-6.901	0	0
8	0.105	1.603	1	-Inf	-6.534	0	NaN
9	-0.080	-2.675	1	2.989	Inf	0	0

Note: coefficients shown are partial regression coefficients

The results presented earlier suggest that the adaptive ridge results for the
anorexia data are unstable, given the difficulties with obtaining lower and upper
bounds for some of the confidence intervals. Thus, it would not be advisable to

interpret them, but rather we may go back to the adaptive lasso approach if we want to use an adaptive regularization parameter for these data.

Grouped Regularization for Dichotomous Logistic Regression

In Chapters 2 and 3 we saw that it is possible to group predictors together to ensure that when regularization is applied, variables within the same group will either all be included in the final model, or none of them will be included. Such grouping is particularly useful in situations where variables are conceptually related to one another, or represent dummy coding for a single categorical variable. In the anorexia example, we can define 4 sets of variables, with emancipation, friends, and school representing relationships, mood, preoccupation, and body representing mental health, sexual attitudes and sexual behavior reflecting sex, and weight standing as a group on its own. We define the groups below using the principles that we described in Chapter 3.

```
group<-c("relationship", "relationship", "relationship",
"mental", "mental", "mental", "sex", "sex", "weight")
```

Next, fit the grouped lasso logistic regression model using the `grpreg` function from the `grpreg` library. The key difference between its application here and in Chapter 3 is that we define the family as `binomial`.

```
dichotomous_grp_lasso.fit<-grpreg(chapter4.x, chapter4.
lr.data$diag2, group, nlambda=100,
    family="binomial", penalty="grLasso")
```

We can view a plot of the estimates by the value of λ.

```
plot(dichotomous_grp_lasso.fit)
```

Based on the results in Figure 4.3, it is clear that for larger values of λ (greater regularization), only a single independent variable had a nonzero coefficient value. Three of the coefficients become nonzero at a λ of approximately 0.05, with additional coefficients departing from 0 at roughly 0.04, and finally, additional coefficients become nonzero at a λ of approximately 0.03.

Selection of the optimal grouped lasso model can be done using by finding the model corresponding to the minimum of an information index such as the BIC. The results appear as follows:

```
select(dichotomous_grp_lasso.fit, "BIC")
$beta
```

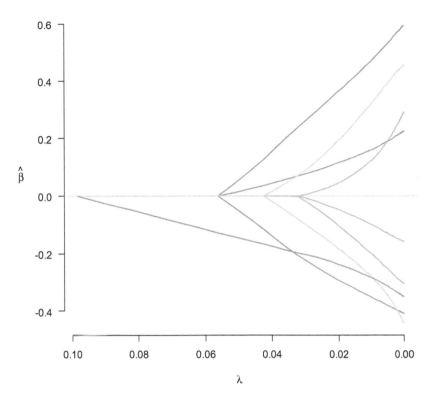

FIGURE 4.3
Coefficient values by λ

```
(Intercept)        eman        frie       school        mood         preo         body         satt
 -0.2127808   0.0000000   0.0000000    0.0000000   0.0000000    0.0000000    0.0000000    0.0000000
       sbeh      weight
  0.0000000   0.0000000
$lambda
[1] 0.09849504
```

The model yielding the lowest BIC included no nonzero coefficients, with a λ
of 0.099. Interestingly, using AIC yields a much different result.

```
select(dichotomous_grp_lasso.fit, "AIC")
$beta
(Intercept)          eman          frie        school          mood           preo           body          satt
 0.01326850   -0.06296362   -0.02835839    0.02383835   0.31770252   -0.26281672    0.09762479   -0.14448910
       sbeh        weight
  0.14350007   -0.22064981
$lambda
[1] 0.02439797
```

There has not been dedicated research to ascertain which, if either, of these information indices is optimal for selecting the grouped lasso model. However, future work may provide greater clarity on this issue.

Logistic Regression for Ordinal Outcome

The previous examples focused on the case where the dependent variable has two categories. It is possible to extend logistic regression to the case where the outcome consists of three or more categories that are ordered. As a way to motivate our discussion of this ordinal logistic regression model, let's return to the anorexia dataset. However, rather than focusing on the dichotomous diagnosis outcome, our dependent variable here will be a counselor's rating of the subjects' family life, which has three levels with 1=most dysfunctional, 2=somewhat dysfunctional, and 3=not dysfunctional. We will use the same set of predictors as we did for binary logistic regression, and which are described in Table 4.1.

A commonly used model for ordinal data such as these is the cumulative logits model, which is as expressed as:

$$\log\text{it}\left[P(Y \le j)\right] = \ln\left(\frac{P(Y \le j)}{1 - P(Y \le j)}\right) \tag{4.12}$$

In this model, there are $J-1$ logits where J is the number of categories in the dependent variable, and Y is the actual outcome value. Essentially, this model compares the likelihood of the outcome variable taking a value of j or lower, versus outcomes larger than j. For the current example there would be 4 separate logits:

$$\ln\left(\frac{p(Y=0)}{p(Y=1) + p(Y=2) + p(Y=3) + p(Y=4)}\right) = \beta_{01} + \beta_1 x$$

$$\ln\left(\frac{p(Y=0) + p(Y=1)}{p(Y=2) + p(Y=3) + p(Y=4)}\right) = \beta_{02} + \beta_1 x$$

$$\ln\left(\frac{p(Y=0) + p(Y=1) + p(Y=2)}{p(Y=3) + p(Y=4)}\right) = \beta_{03} + \beta_1 x \tag{4.13}$$

$$\ln\left(\frac{p(Y=0) + p(Y=1) + p(Y=2) + p(Y=3)}{p(Y=4)}\right) = \beta_{04} + \beta_1 x$$

In the cumulative logits model, there is a single slope relating the independent variable to the ordinal response, and each logit has a unique intercept. In order for a single slope to apply across all logits we must make the proportional odds assumption, which states that this slope is identical across logits. If this assumption is violated, we will need to use an alternative model that doesn't make this assumption, such as the multinomial logit model.

In order to fit a cumulative logits regression model to our dataset, we can use the `polr` function from the MASS library, which is typically loaded each time you open R. To be sure that it loads properly, you can use `library(MASS)`. The following commands will first fit the model, and then present the parameter estimates:

```
family.model<-polr(as.factor(chapter4.lr.data$fami)~eman+f
rie+school+mood+preo+body+satt+sbeh+weight, data=chapter4.
lr.data, method="logistic")

summary(family.model)

Re-fitting to get Hessian

Call:
polr(formula = as.factor(chapter4.lr.data$fami) ~ eman +
frie +
    school + mood + preo + body + satt + sbeh + weight,
data = chapter4.lr.data,
    method = "logistic")

Coefficients:
            Value Std. Error  t value
eman     0.895468     0.2443  3.66478
frie     0.541804     0.2030  2.66950
school   0.058899     0.2842  0.20724
mood     0.017287     0.2243  0.07705
preo     0.163143     0.2597  0.62810
body     0.441091     0.2407  1.83220
satt    -0.007045     0.2838 -0.02482
sbeh     0.077116     0.2660  0.28993
weight   0.152796     0.1265  1.20782

Intercepts:
     Value   Std. Error t value
1|2  3.5042  0.8097      4.3280
2|3  5.4272  0.8638      6.2830

Residual Deviance: 366.1703
AIC: 388.1703
```

The resulting output includes the parameter estimate, standard error, and the ratio of the two, which is in the `t value` column. The intercepts appear below the coefficients, followed by the deviance and AIC statistics. If we want hypothesis tests associated with these estimates, we can use the Anova command from the car library.

```
library(car)
Anova(family.model)
Analysis of Deviance Table (Type II tests)

Response: as.factor(chapter4.lr.data$fami)
          LR Chisq Df Pr(>Chisq)
eman      13.9326   1  0.0001895 ***
frie       7.2253   1  0.0071883 **
school     0.0429   1  0.8359374
mood       0.0059   1  0.9385847
preo       0.3938   1  0.5303005
body       3.3781   1  0.0660692 .
satt       0.0006   1  0.9801904
sbeh       0.0839   1  0.7720363
weight     1.4541   1  0.2278674
---
Signif. codes:  0 '***' 0.001 '**' 0.01 '*' 0.05 '.' 0.1 ' ' 1
```

These test results show that the coefficients for eman and frie were statistically significantly different from 0. Combining this finding with the coefficients, we would conclude that individuals with larger eman scores (greater emancipation) and frie scores (more positive relationships with friends) were more likely to have higher fami scores (more positive relationships with family). None of the other variables were found to be statistically significant.

Finally, we need to investigate the proportional odds assumption. This can be done in R using the poTest command. We simply need to provide the model output object and will then obtain tests of the null hypothesis that the proportional odds assumption holds for each variable.

```
poTest(family.model)

Tests for Proportional Odds
polr(formula = as.factor(chapter4.lr.data$fami) ~ eman + frie +
    school + mood + preo + body + satt + sbeh + weight,
data = chapter4.lr.data,
    method = "logistic")

           b[polr]     b[>1]     b[>2] Chisquare df Pr(>Chisq)
Overall                                     8.79  9      0.457
eman       0.89547   1.06482   0.79790      0.56  1      0.453
frie       0.54180   0.47523   0.55074      0.08  1      0.781
school     0.05890  -0.12923   0.00950      0.11  1      0.745
mood       0.01729  -0.06539  -0.03475      0.01  1      0.922
preo       0.16314   0.58269   0.01894      2.28  1      0.131
body       0.44109  -0.00919   0.69384      4.29  1      0.038 *
```

```
satt    -0.00704   0.32033  -0.07350      0.96   1      0.326
sbeh     0.07712   0.23311   0.01074      0.31   1      0.577
weight   0.15280   0.27751   0.17332      0.32   1      0.574
---
Signif. codes:   0 '***' 0.001 '**' 0.01 '*' 0.05 '.' 0.1 ' ' 1
```

Based on these results, it appears that the proportional odds assumption may not hold for the independent variable body. Thus, we may want to consider using a semiparametric ordinal regression model, which allows for the relaxation of this assumption for specific predictors. This model is beyond the scope of our work here, but the interested reader is referred to Agresti (2013) for a further discussion of this model.

Regularized Ordinal Logistic Regression

In order to fit a regularized cumulative logits model, we will use the ordinalNet library in R. As with glmnet, we must first place the set of independent variables to be used in the analysis into a matrix.

```
chapter4.x<-as.matrix(chapter4.lr.data[,c("eman","frie","sc
hool","mood","preo","body","satt","sbeh","weight")])
```

We will use the ordinalNet function to fit the regularized cumulative logits regression models. For the lasso model, we need to set alpha=1. By default, ordinalNet explores 20 λ values, although we can set both the number and range of λ's if we so choose. In this example, we will rely on the default search options. We must specify the matrix of predictors as well as provide a factor version of the dependent variable, which can be done using the as.factor function. We define the model as a cumulative logits regression through the family and link subcommands, and the lasso estimator is used when we set alpha=1. Finally, we assume the parallel odds assumption by using parallelTerms=TRUE, nonparallelTerms=FALSE. If we specify parallelTerms=FALSE, nonparallelTerms=TRUE, then we do not force the parallel odds assumption to hold, and all coefficients are regularized toward 0. Finally, if we use parallelTerms=TRUE, nonparallelTerms=TRUE, we are requesting a semiparametric model in which the coefficients are regularized toward a common value, but are not forced to be a common value. Following the model specification, we print the values of λ that were explored, along with the number of coefficients that were not 0 for that penalty, and the log-likelihood, deviance, AIC, and BIC values associated with each λ.

```
family.lasso <- ordinalNet(chapter4.x, as.factor(chapter4.
lr.data$fami), family="cumulative", link="logit",
```

```
parallelTerms=TRUE, nonparallelTerms=FALSE, alpha=1)
summary(family.lasso)
      lambdaVals nNonzero   loglik      devPct       aic      bic
1   0.225677829        2 -218.4974 1.110223e-16 440.9948 447.7546
2   0.177102927        3 -210.1383 3.825719e-02 426.2766 436.4163
3   0.138983287        5 -201.7256 7.675978e-02 413.4512 430.3507
4   0.109068521        6 -195.4295 1.055752e-01 402.8590 423.1384
5   0.085592609        7 -191.1246 1.252776e-01 396.2492 419.9084
6   0.067169653        7 -188.2381 1.384882e-01 390.4762 414.1355
7   0.052712055        7 -186.3924 1.469355e-01 386.7848 410.4441
8   0.041366311        8 -185.1972 1.524057e-01 386.3943 413.4335
9   0.032462625        8 -184.4310 1.559124e-01 384.8619 411.9011
10  0.025475369        8 -183.9449 1.581369e-01 383.8898 410.9290
11  0.019992049        9 -183.6298 1.595792e-01 385.2595 415.6786
12  0.015688960        9 -183.4263 1.605104e-01 384.8526 415.2717
13  0.012312068        9 -183.2988 1.610941e-01 384.5975 415.0166
14  0.009662018        9 -183.2191 1.614588e-01 384.4382 414.8572
15  0.007582365        9 -183.1694 1.616861e-01 384.3389 414.7579
16  0.005950337        9 -183.1386 1.618273e-01 384.2771 414.6962
17  0.004669586        9 -183.1194 1.619150e-01 384.2388 414.6579
18  0.003664504       10 -183.1067 1.619733e-01 386.2133 420.0123
19  0.002875756       10 -183.0986 1.620104e-01 386.1971 419.9961
20  0.002256778       10 -183.0936 1.620333e-01 386.1871 419.9861
```

The coefficients associated with the smallest BIC and AIC values appear as follows, by using the criteria subcommand with the coef function:

```
coef(family.lasso, criteria="bic")

(Intercept):1 (Intercept):2          eman         frie       school         mood
   1.86960628    3.61246813   -0.72463074  -0.35868114   0.00000000   0.00000000
         preo          body          satt         sbeh       weight
  -0.09704541   -0.31271178    0.00000000   0.00000000  -0.07370352

coef(family.lasso, criteria="aic")

(Intercept):1 (Intercept):2          eman         frie       school         mood
   2.63881786    4.46665517   -0.81508643  -0.45369733   0.00000000   0.00000000
         preo          body          satt         sbeh       weight
  -0.12770152   -0.37378119    0.00000000  -0.04146776  -0.11782670
```

The minimum BIC was associated with a l of 0.052712055 (row 7), whereas the minimum AIC had l of 0.025475369 (row 10). For the BIC solution, eman, frie, preo, body, and weight all had non-zero coefficients. For the optimal AIC solution, eman, frie, preo, body, sbeh, and weight had non-zero coefficients. An important point to note is that the cumulative logits used by ordinalNet is the reverse of the standard logits, thereby leading to the reversal in signs that we see here. With that difference in mind, the lasso estimates reflect a similar finding with higher values of eman and frie being associated with a more positive family relationship. There is currently not an option to conduct hypothesis testing for the coefficients provided by the ordinalNet library.

The ridge regression estimator can be fit very similarly to the lasso, with the only difference being that we set alpha=0.

```
family.ridge <- ordinalNet(chapter4.x, as.factor(chapter4.lr.data$fami),
family="cumulative", link="logit",
+ parallelTerms=TRUE, nonparallelTerms=FALSE, alpha=0)

summary(family.ridge)
```

	lambdaVals	nNonzero	loglik	devPct	aic	bic
1	22.5677829	11	-216.3148	0.009989338	454.6295	491.8084
2	17.7102927	11	-215.7572	0.012541244	453.5144	490.6932
3	13.8983287	11	-215.0704	0.015684623	452.1407	489.3196
4	10.9068521	11	-214.2317	0.019523052	450.4634	487.6422
5	8.5592609	11	-213.2185	0.024159955	448.4371	485.6159
6	6.7169653	11	-212.0107	0.029688039	446.0213	483.2002
7	5.2712055	11	-210.5934	0.036174336	443.1868	480.3657
8	4.1366311	11	-208.9616	0.043642879	439.9231	477.1020
9	3.2462625	11	-207.1242	0.052051876	436.2494	473.4273
10	2.5475369	11	-205.1070	0.061283899	432.2141	469.3930
11	1.9992049	11	-202.9538	0.071138608	427.9076	465.0865
12	1.5688960	11	-200.7237	0.081344969	423.4475	460.6264
13	1.2312068	11	-198.4854	0.091589459	418.9707	456.1496
14	0.9662018	11	-196.3078	0.101555412	414.6156	451.7945
15	0.7582365	11	-194.2523	0.110962856	410.5046	447.6835
16	0.5950337	11	-192.3656	0.119597847	406.7312	443.9101
17	0.4669586	11	-190.6772	0.127325079	403.3544	440.5333
18	0.3664504	11	-189.2003	0.134084381	400.4006	437.5795
19	0.2875756	11	-187.9347	0.139876738	397.8694	435.0483
20	0.2256778	11	-186.8706	0.144746750	395.7412	432.9201

```
coef(family.ridge, criteria="bic")
```

(Intercept):1	(Intercept):2	eman	frie	school	mood
2.15201648	3.89515062	-0.43579431	-0.30355597	-0.09938199	-0.02179388
preo	body	satt	sbeh	weight	
-0.21060020	-0.27060826	-0.09300339	-0.16671386	-0.10375000	

```
coef(family.ridge, criteria="aic")
```

(Intercept):1	(Intercept):2	eman	frie	school	mood
2.15201648	3.89515062	-0.43579431	-0.30355597	-0.09938199	-0.02179388
preo	body	satt	sbeh	weight	
-0.21060020	-0.27060826	-0.09300339	-0.16671386	-0.10375000	

The optimal λ was the same for both AIC and BIC. Note that the coefficients for the two most important variables (eman and frie) were much more severely regularized by the ridge estimator, as opposed to the lasso.

The basic conclusion regarding relationships among the variables would be essentially the same for the two estimators.

Finally, we can apply the elastic net estimator using ordinalNet. There is not an automated function for doing so, therefore we will need to manually change the values of α, and then compare the AIC/BIC values in order to determine the optimal solution. Below is the code for fitting the elastic net with alpha=0.1. The algorithm will then search a range of λ values with α=0.1. The basic function call is nearly identical to that for the lasso and ridge estimators, with the only difference being the value of alpha. The additional commands identify the summary row with the minimum AIC, then print the coefficients associated with this value, and the AIC itself. The same information is then obtained for the optimal BIC value.

```
family.net1 <- ordinalNet(chapter4.x, as.factor(chapter4.
lr.data$fami), family="cumulative", link="logit",
parallelTerms=TRUE, nonparallelTerms=FALSE, alpha=0.1)

summary(family.net1)
```

	lambdaVals	nNonzero	loglik	devPct	aic	bic
1	2.25677829	2	-218.4974	1.110223e-16	440.9948	447.7546
2	1.77102927	4	-217.1126	6.338127e-03	442.2251	455.7447
3	1.38983287	8	-213.2648	2.394825e-02	442.5296	469.5687
4	1.09068521	10	-208.5442	4.555281e-02	437.0885	470.8875
5	0.85592609	10	-203.9634	6.651824e-02	427.9267	461.7257
6	0.67169653	10	-199.8933	8.514548e-02	419.7867	453.5857
7	0.52712055	10	-196.3873	1.011918e-01	412.7745	446.5735
8	0.41366311	10	-193.4469	1.146489e-01	406.8938	440.6928
9	0.32462625	10	-191.0372	1.256777e-01	402.0743	435.8733
10	0.25475369	10	-189.1011	1.345387e-01	398.2021	432.0011
11	0.19992049	10	-187.5725	1.415343e-01	395.1451	428.9440
12	0.15688960	10	-186.3851	1.469688e-01	392.7702	426.5692
13	0.12312068	10	-185.4770	1.511252e-01	390.9539	424.7529
14	0.09662018	10	-184.7933	1.542540e-01	389.5866	423.3856
15	0.07582365	10	-184.2870	1.565714e-01	388.5739	422.3729
16	0.05950337	10	-183.9183	1.582586e-01	387.8366	421.6356
17	0.04669586	10	-183.6546	1.594656e-01	387.3092	421.1082
18	0.03664504	10	-183.4692	1.603140e-01	386.9384	420.7374
19	0.02875756	10	-183.3412	1.609000e-01	386.6823	420.4813
20	0.02256778	10	-183.2541	1.612983e-01	386.5083	420.3073

```
family.net1.min.aic<-which.min(family.net1$aic)
family.net1$coefs[family.net1.min.aic,]
```

(Intercept):1	(Intercept):2	eman	frie	school	mood
3.20461690	5.08847560	-0.78953257	-0.48931376	-0.06488390	0.00000000
preo	body	satt	sbeh	weight	
-0.19342308	-0.39959025	-0.02220386	-0.10471854	-0.14055449	

```
family.net1$aic[family.net1.min.aic]
```

```
[1] 386.5083

family.net1.min.bic<-which.min(family.net1$bic)

family.net1$coefs[family.net1.min.bic,]
```

(Intercept):1	(Intercept):2	eman	frie	school	mood
3.20461690	5.08847560	-0.78953257	-0.48931376	-0.06488390	0.00000000
preo	body	satt	sbeh	weight	
-0.19342308	-0.39959025	-0.02220386	-0.10471854	-0.14055449	

```
family.net1$bic[family.net1.min.bic]

[1] 420.3073
```

In order to identify an optimal elastic net result for the data, we can use the same set of commands with values of alpha ranging from 0.1 to 0.9. The resulting AIC and BIC results appear in Table 4.2.

The minimum values for both AIC and BIC are associated with α of 0.9. Thus, we will examine the coefficients for the model associated with this value, along with its optimal λ, which is 0.028305965.

```
coef(family.net9, criteria="bic")
```

(Intercept):1	(Intercept):2	eman	frie	school	mood
2.61372316	4.43794516	-0.80331282	-0.44941227	0.00000000	0.00000000
preo	body	satt	sbeh	weight	
-0.13124456	-0.36883697	0.00000000	-0.04776727	-0.11727178	

```
coef(family.net9, criteria="aic")
```

(Intercept):1	(Intercept):2	eman	frie	school	mood
2.61372316	4.43794516	-0.80331282	-0.44941227	0.00000000	0.00000000
preo	body	satt	sbeh	weight	
-0.13124456	-0.36883697	0.00000000	-0.04776727	-0.11727178	

TABLE 4.2

Optimal AIC and BIC values for each value of alpha

Alpha	AIC	BIC
0.1	386.5083	420.3073
0.2	386.4334	420.2323
0.3	386.3143	420.1133
0.4	386.2088	420.0078
0.5	384.2886	414.7076
0.6	384.2689	414.688
0.7	384.2567	414.3853
0.8	384.2486	412.2739
0.9	383.9977	411.0369

The degree of regularization is relatively less severe than was the case for either the ridge or lasso estimators, with coefficients for school, mood, and satt all being set to 0.

One final point to note here is that we selected only 10 values for alpha with this example. However, it is possible (indeed likely) that a larger range of such values would allow for a more precise identification of the optimal elastic net parameter estimates. Therefore, although somewhat tedious, it may be useful for the data analyst to investigate a larger range of alpha values than was used for this example.

Regression Models for Count Data

Poisson Regression

To this point, we have been focused on outcome variables of a categorical nature, such as the diagnosis of a particular malady or the degree to which family relationships are positive. Another type of data that does not fit nicely into the standard models assuming normally distributed errors involves counts or rates of some outcome, particularly of rare events. Such variables often follow the Poisson distribution, a major property of which is that the mean is equal to the variance. It is clear that if the outcome variable is a count, its lower bound must be be 0; i.e., one cannot have negative counts. This presents a problem to researchers applying the standard linear regression model, as it may produce predicted values of the outcome that are less than 0, and thus are nonsensical. In order to deal with this potential difficulty, Poisson regression was developed. This approach to dealing with count data rests upon the application of the log to the outcome variable, thereby overcoming the problem of negative predicted counts, since the log of the outcome can take any real number value. Thus, when dealing with the Poisson distribution in the form of counts, we will use the log as the link function in fitting the Poisson regression model:

$$\ln(Y) = \beta_0 + \beta_1 x \tag{4.14}$$

In all other respects, the Poisson model is similar to other regression models in that the relationship between the independent and dependent variables is expressed through the slope, β_1. And again, the assumption underlying the Poisson model is that the mean is equal to the variance. This assumption is typically expressed by stating that the overdispersion parameter, $\phi=1$. The ϕ parameter appears in the Poisson distribution density function and thus is a key component in the fitting function used to determine the optimal model parameter estimates in maximum likelihood. A thorough review of this fitting function is beyond the scope of this book. Interested readers are referred to Agresti (2013) for a complete presentation of this issue.

Before we discuss the Poisson regression model using regularized estimators, let's first use the standard approach for Poisson regression, which can be done with the `glm` function in R. Recall that `glm` is part of the MASS library, which typically loads automatically when we open R. The example that we will use involves a sample of 133 high school students, for each of whom was recorded the number of citations given by the school for some type of misbehavior over the course of a school year. In addition, for each student, standardized scores on each of nine measures of executive functioning were also collected. The school psychologist conducting this research is interested in ascertaining which of these executive functioning scores are associated with the number of citations.

We can examine the distribution of number of citations using a histogram.

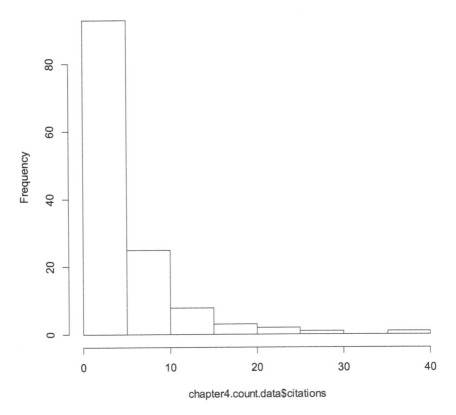

FIGURE 4.4
Frequency of citation number

Generally speaking, students received 0 or only a few citations over the course of the school year. However, a small number of students had more than 10 citations, with one individual having nearly 40.

The Poisson regression model can be fit, and the results can be obtained using the following commands. We will save the output in the object `chapter4.poisson.model`. The function call requires the regression model that we want to fit, the dataframe name, and the link function, which reflects the distribution of the dependent variable.

```
chapter4.poisson.model<-glm(citations~ef1+ef2+ef3+ef4+ef5+ef6+ef7+
ef8+ef9, data=chapter4.count.data, family=c("poisson"))

summary(chapter4.poisson.model)

Call:
glm(formula = citations ~ ef1 + ef2 + ef3 + ef4 + ef5 + ef6 +
    ef7 + ef8 + ef9, family = c("poisson"), data = chapter4.count.data)

Deviance Residuals:
    Min       1Q    Median       3Q      Max
-2.2527  -0.8864   -0.2357   0.5452   2.5431

Coefficients:
             Estimate Std. Error z value Pr(>|z|)
(Intercept)   0.99471    0.06199  16.046  < 2e-16 ***
ef1          -0.53657    0.04336 -12.375  < 2e-16 ***
ef2          -0.01345    0.03592  -0.374    0.708
ef3          -0.24659    0.04796  -5.142 2.72e-07 ***
ef4           0.03666    0.04070   0.901    0.368
ef5          -0.03070    0.04577  -0.671    0.502
ef6          -0.90856    0.04940 -18.390  < 2e-16 ***
ef7           0.06169    0.04036   1.529    0.126
ef8          -0.65656    0.04349 -15.098  < 2e-16 ***
ef9           0.01129    0.04195   0.269    0.788
---
Signif. codes:  0 '***' 0.001 '**' 0.01 '*' 0.05 '.' 0.1 ' ' 1

(Dispersion parameter for poisson family taken to be 1)

    Null deviance: 784.18  on 132  degrees of freedom
Residual deviance: 131.25  on 123  degrees of freedom
AIC: 503.32
```

Given these results, we would conclude that the scores on tests 1, 3, 6, and 8 were all statistically significantly related to the number of citations issued by the school. Each of the significant coefficients was negative, meaning that higher executive functioning scores were associated with fewer citations. In addition, five of the nine predictors were not found to be related to the number of citations.

Regularized Count Regression

The regularization approaches that we have discussed throughout this chapter can be applied to count regression models. The negative of the log-likelihood for the lasso penalty in this context is

$$-\frac{1}{N}\sum_{i=1}^{N}\left\{y_i\left(\beta_0+\beta^T x_i\right)-e^{\beta_0+\beta^T x_i}\right\}+\lambda\|\beta\|_1.$$ (4.15)

where
    N = Total sample size
    y_i = Response variable value for subject i
    β_0 = Intercept
    β^T = Vector of coefficients for independent variables
    x_i = Vector independent variable values for subject i
    β = Model coefficients, excluding the intercept
    λ = Regularization parameter

The framework underlying the lasso estimator with Poisson regression is essentially the same as for other models that we examine in this book. Note that the intercept is typically not included in the regularization process, meaning that it will not be penalized. The ridge estimator takes the form

$$-\frac{1}{N}\sum_{i=1}^{N}\left\{y_i\left(\beta_0+\beta^T x_i\right)-e^{\beta_0+\beta^T x_i}\right\}+\lambda\|\beta\|_2^2$$ (4.16)

Finally, the elastic net for the Poisson regression model is

$$-\frac{1}{N}\sum_{i=1}^{N}\left\{y_i\left(\beta_0+\beta^T x_i\right)-e^{\beta_0+\beta^T x_i}\right\}+\lambda\left[\frac{1}{2}(1-\alpha)\|\beta\|_2^2+\alpha\|\beta\|_1\right]$$ (4.17)

where
    $\alpha\in[0,1]$

Now that we have seen the regularization estimators for the Poisson regression model, let's apply them to the citation data using R. Given the relatively large number of independent variables coupled with a sample of 133 individuals, it is worth applying the regularization techniques. We can obtain these estimates using the glmnet library, as we did for the logistic regression model. One important point to note is that the tools we used for inference with the normal-based and dichotomous logistic regression models is not available for use with Poisson regression models. The commands to obtain the lasso estimates of the Poisson regression model parameters appear below. We will use the familiar cv.glmnet function, and thus are familiar with the structure of the function call. As we have seen in this chapter (as well as Chapter 3), we can plot the regularization tuning parameter by MSE, and can also obtain the optimal value (lambda.min) for our model.

```
chapter4.poisson.x<-as.matrix(chapter4.count.data[,c("ef1",
"ef2","ef3","ef4","ef5","ef6","ef7","ef8","ef9")])

poisson.lasso.cv <- cv.glmnet(chapter4.poisson.x, chapter4.
count.data$citations, alpha = 1, family = "poisson",
nfolds=10,type.measure="mse")

plot(poisson.lasso.cv)

poisson.lasso.cv$lambda.min
[1] 0.01157391

poisson.lasso.cv$lambda.1se
[1] 0.1566005
```

Based on the MSE from the 10-fold cross-validation, the optimal tuning parameter value is 0.0116. Because this value is so close to 0, we know that there is relatively little regularization being done by the lasso estimator.

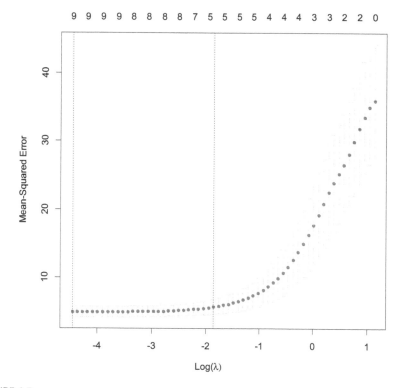

FIGURE 4.5
Regularization tuning parameter by MSE for Poisson regression model with lasso penalty

Now that we've identified the optimal value of λ, we can fit the optimal Poisson regression model, and then obtain the coefficients for the result.

```
poisson.lasso.optimal <- glmnet(chapter4.poisson.x,
chapter4.count.data$citations, alpha = 0, family =
"poisson", lambda = poisson.lasso.cv$lambda.min)

coef(poisson.lasso.optimal)
10 x 1 sparse Matrix of class "dgCMatrix"
                      s0
(Intercept)   0.99978199
ef1          -0.53333304
ef2          -0.01374096
ef3          -0.24532945
ef4           0.03644238
ef5          -0.03062337
ef6          -0.90397697
ef7           0.06178717
ef8          -0.65410545
ef9           0.01086470
```

The coefficients for the lasso estimator are very similar to those obtained for the standard Poisson regression model, which we anticipated given the value of the regularization parameter.

We can fit the ridge estimator in much the way we did the lasso.

```
poisson.ridge.cv <- cv.glmnet(chapter4.poisson.x, chapter4.
count.data$citations, alpha = 0, family = "poisson",
nfolds=10, type.measure="mse")

plot(poisson.ridge.cv)

poisson.ridge.cv$lambda.min
[1] 0.3074132

poisson.ridge.cv$lambda.1se
[1] 0.7101643
```

For the ridge estimator, the value of λ is somewhat larger than was the case for the lasso, indicating that more smoothing took place. The final parameter estimates appear as follows:

```
poisson.ridge.optimal <- glmnet(chapter4.poisson.x,
chapter4.count.data$citations, alpha = 0, family =
"poisson", lambda = poisson.ridge.cv$lambda.min)

coef(poisson.ridge.optimal)

10 x 1 sparse Matrix of class "dgCMatrix"
                      s0
(Intercept)   1.102628077
```

```
ef1            -0.465375728
ef2            -0.019175014
ef3            -0.219245386
ef4             0.032413891
ef5            -0.029178350
ef6            -0.806983754
ef7             0.062315209
ef8            -0.600725664
ef9             0.001699325
```

The ridge coefficients are somewhat smaller than those of the lasso, reflecting the larger value of λ associated with it, vis-à-vis the lasso. In terms of our final conclusions regarding the relationships between the executive functioning scores and the number of citations, however, these results do not differ substantially from those of either the standard Poisson regression, or the lasso.

Finally, we can fit the model linking number of citations to the measures of executive functioning using the elastic net using the set of commands from

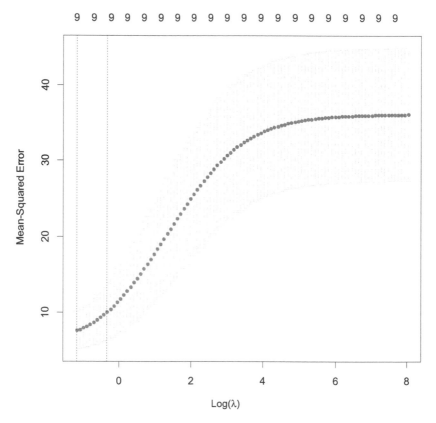

FIGURE 4.6
Regularization tuning parameter by MSE for Poisson regression model with ridge penalty

the `caret` function that we demonstrated in Chapter 3. Following are the commands for the elastic net with the Poisson regression:

```
poisson.net.model <- train(
  citations~ef1+ef2+ef3+ef4+ef5+ef6+ef7+ef8+ef9, data =
chapter4.count.data, method = "glmnet",
  trControl = trainControl("cv", number = 10),
family="poisson",
  tuneLength = 10
)
```

As described in Chapter 3, we need to provide the function with the model equation, the dataset, the R function to use (`glmnet`), the type of model check (10 fold cross-validation in this case), the dependent variable distribution (Poisson), and the number of values for each tuning parameter. Thus, we will have 10 values for α, and 10 for λ, yielding 100 total combinations.

Once we've fit the training model, we can examine the optimal tuning parameters depending upon which function we want to optimize. As we noted in Chapter 3, the default statistic used to obtain this optimal value is RMSE. The optimal values of α and λ based on RMSE are

```
poisson.net.model$bestTune
    alpha      lambda
76     1 0.09344812
```

With regard to the MAE, the optimal regularization parameters are

```
poisson.net.model$results[which.min(poisson.net.model$results$MAE),]
    alpha   lambda    RMSE Rsquared      MAE  RMSESD RsquaredSD    MAESD
15    0.2 1.152072 6.179323 0.643673 3.622878 1.825693 0.09438889 0.7509601
```

Thus, in order to optimize MAE, we would set α=0.2 and λ=1.152072. We will examine the coefficients for both sets of regularization parameters as follows:

```
poisson.net.optimal.mae <- glmnet(chapter4.poisson.x,
chapter4.count.data$citations, alpha = 0.2, family =
"poisson",lambda = 1.152072)

coef(poisson.net.optimal.mae)

10 x 1 sparse Matrix of class "dgCMatrix"
                   s0
(Intercept)   1.30924526
ef1          -0.31377616
ef2               .
ef3          -0.14053377
ef4               .
ef5               .
```

```
ef6            -0.60278852
ef7             0.02318912
ef8            -0.46989885
ef9                 .
```

From these results, we can see that ef1, ef3, ef6, ef7, and ef8 were not regularized to 0. Although the magnitudes of the coefficients differ from those produced by the Ridge and lasso estimators, qualitatively they tell the same general story, with ef1, ef3, ef6, and ef8 all having negative associations with the number of disciplinary citations.

Cox Proportional Hazards Model

Survival analysis, also sometimes referred to as time until event data, is used when the outcome variable is the time until a particular event of interest occurs. Such events might include death, onset of illness, scoring a passing mark on an exam, or promotion in one's job. What makes this type of data unique is the possibility that an individual will leave the sample at some point prior to the event occurring. For example, consider the case where researchers collect data on the time until onset of cardiovascular disease in adults over 50 years of age. Some members of the sample will develop heart disease during the course of the study, yielding a measure for the time until the event. However, for others, this will not be the case, either due to death from other causes, their withdrawal from the study, or the completion of the study prior to their developing heart disease. These individuals are said to be censored because the event did not occur for them during the course of the study, but we cannot say that the event will never occur. For these subjects, the time until censoring occurred is recorded, and they are identified as having been censored.

Typically, a first step in the analysis of survival data is an examination of the pattern of event occurrence over time. This can be done in a variety of ways, with perhaps the most popular being Kaplan-Meier (KM) plots, which graph the time until the event on the x-axis, and the probability of the event *not* occurring on the y-axis. This plot basically graphs the survival function, and we will take a look at it with an example later. The KM survival function is estimated as:

$$s(t) = \prod_{j=1}^{T} \left[1 - \frac{d_j}{n_j} \right]$$

(4.18)

where

    d_j = Number for whom the event occurred at time t_j.
    n_j = Number at risk at time t_j.

It is important to note here that n_j includes only those individuals who remain in the sample at time t_j; i.e., those for whom the event had already occurred, and those who were censored are excluded. Survival functions can be compared between two groups (e.g., females and males) using a test statistic such as the log-rank test:

$$Log - Rank = \sum_{j=1}^{t} \left(d_{1j} - e_{1j} \right) \qquad (4.19)$$

where

d_{1j} = Observed number of events for group 1 at time j.
e_{1j} = Expected number of events for group 1 at time j, if H_0 is true.

$$e_{1j} = \frac{\left(n_{1j} \right)\left(d_j \right)}{\left(n_j \right)}$$

n_{1j} = Number in group 1 at time j
d_j = Total number of individuals for whom the event occurred at time j
n_j = Total number of individuals in the sample at time j

The null hypothesis is that the survival function is the same for the two groups.

An alternative way to view the data is with the hazard function. Conceptually, we can think of the hazard as the instantaneous likelihood of the event of interest occurring at any point in time. From a statistical perspective, we can view the hazard as the probability of the event occurring during a small window of time. As was true of the survival function, hazard at a specific time, t, is conditional only on those surviving until that time.

In addition to the relatively simple KM approach outlined earlier, it is also possible to more formally model the survival (or hazard) functions, and to relate them with one or more independent variables. Some of these models are parametric in nature, assuming that the survival function is of a particular form, such as the exponential distribution. An alternative approach is the Cox proportional hazards model, which does not make any assumptions about the underlying distribution of the time until the event of interest. The only assumption underlying the Cox model is that the hazard rate for different individuals in the sample has a fixed ratio (is constant) across all time periods. The Cox model allows for the inclusion of multiple independent variables to serve as predictors of the hazard rate. This relationship can be expressed as

$$ln\left(h_i\left(t \right) \right) = ln\left(\lambda_0\left(t \right) \right) + \beta_1 x_{1i} + \beta_2 x_{2i} + \cdots + \beta_j x_{ji} . \qquad (4.20)$$

where

$h_i\left(t \right)$ = Hazard rate for subject i at time t
$\lambda_0\left(t \right)$ = Baseline hazard function

β_j = Coefficient relating variable j tohe hazard rate
x_{ji} = Value on independent variable j for subject i

The model parameters can be estimated using partial maximum likelihood estimation. The fit of the model can be assessed using the concordance statistic, which measures how closely predictions obtained from the estimated model correspond to the actual data. Correspondence ranges between 0 and 1, with higher values indicating better model fit. Concordance is calculated using the following steps:

1. Calculate predicted survival time for each observation.
2. For every possible pair of observations compare the predicted and actual survival times.
3. A pair is counted as concordant if the individual with the higher predicted survival time also has the higher observed survival time.
4. Ties can be counted as concordant, discordant, or as 0.5 concordant.
5. Divide the number of concordant pairs by the total number of pairs in the sample.

In order to demonstrate the fitting of a Cox proportional hazards model, we will take data from a sample of patients who were being treated for colon cancer. The outcome of interest is the time until a patient has a recurrence of cancer over the follow-up period after treatment. A variety of variables (listed in Table 4.3) were measured for each of the patients. These data are available through the `survival` library in R.

Prior to actually obtaining the Kaplan–Meier survival estimates, we need to load the appropriate R libraries and then access the data.

TABLE 4.3

Independent variables measured from heart surgery patients

Variable name	Description
Sex	1=Male
Age	Years
Obstruct	Tumor obstructed colon; 1=Yes
Perfor	Colon was perforated; 1=Yes
Adhere	Colon adhered to nearby organs; 1=Yes
Nodes	Number of lymph nodes to which cancer spread
Time	Days until event or censoring
Status	Censored; 1=Yes
Differ	Differentiation of tumor; 1=well, 2=moderate, 3=poor
Surg	Time from surgery to registration; 0=short, 1=long

```
library(survival)
library(ggplot2)
library(survMisc)
library(car)
library(leaps)
library(glmnet)
library(selectiveInference)
library(caret)
library(DescTools)
library(grpreg)
```

The data of interest is part of the `survival` library, and can be accessed using the following command:

```
data("cancer")
```

Next, we need to pull out the subset of observations associated with the recurrence of cancer, which is reflected in the variable `etype`. In addition to recurrence, the dataset also includes information about how much time passed before death (or censoring) for each individual in the sample. The following command creates a new dataframe that includes the recurrence information and then removes individuals who have missing data on any of the variables.

```
colon_recur<-colon[which(colon$etype==1),]
colon_recur<-na.omit(colon_recur)
```

We can then attach the new dataframe, which we will be using going forward.

```
attach(colon_recur)
```

In order to fit any of the survival models to the data, we must create an object that reflects both the time until the event (or censoring), as well as the event code itself (censored or event). This new object, created below using the `Surv` function and named `colon.km.eqn`, will serve as the dependent variable in subsequent analyses.

```
colon.km.eqn<-Surv(colon_recur$time, colon_recur$status)
```

We are now ready to examine the Kaplan-Meier (KM) estimate of the survival function for our dataset. The KM estimate presents the probability that an individual will not experience the event (reoccurrence of cancer) by time (number of days). Fitting a KM model to the data is usually a good first step when investigating survival data. It provides us with a simple description of the likelihood of the event occurring over time and serves the same purpose as descriptive statistics (e.g., mean, standard deviation, correlation coefficient) do in other analyses. We can obtain the KM estimate using the `survfit` function from the `survival` library. Once we have fit the simple model (first command below), we can plot the resulting survival curve, along with its 95% confidence interval.

```
colon.km.fit<-survfit(colon.km.eqn~1, data=colon_recur)

plot(colon.km.fit, xlab="Days", ylab="Probability that
cancer does not reoccur", main="Kaplan-Meier survival
curve")
```

In this example, survival equates to not having the colon cancer return. With that fact in mind, this plot reveals that there is a steady reoccurrence of cancer in the sample beginning almost immediately after treatment. The probability of having the cancer return begins to level off after approximately 600 days from the end of the initial treatment. By 1000 days post-treatment

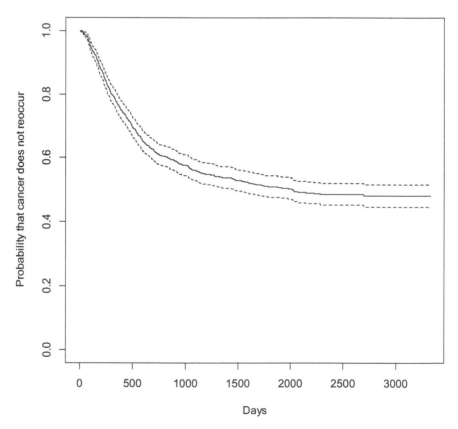

FIGURE 4.7
Kaplan–Meier survival curves for cancer reoccurrence

the probability of *not* having cancer reoccur stabilizes between approximately 50% and 60%, based on the 95% confidence interval.

If we would rather plot the hazard of having cancer reoccur, we can do so by setting `fun="cumhaz"` when we plot the KM results. Also, note that for this plot we set the lower bound of the y-axis to be 0 and the upper bound to be 1.

```
plot(colon.km.fit, xlab="Days", ylim=c(0,1),
 ylab="Probability that cancer does reoccur", main="Kaplan-
Meier hazard curve", fun="cumhaz")
```

The story told by the hazard function is simply the reverse of what we saw with the survival function plot: The probability of cancer reoccurring rose

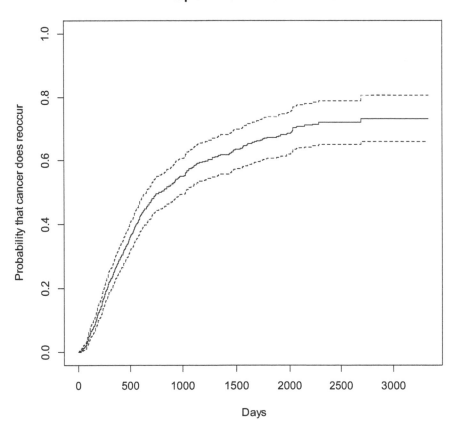

FIGURE 4.8
Kaplan–Meier hazard curves for cancer reoccurrence

at a steady rate for the first 600 days or so after treatment, and then began to level off before stabilizing after approximately 1000 days.

In addition to the overall survival function, it is also possible to obtain separate functions for different groups within the sample. For example, we may want to compare the probability of cancer not reoccurring for those whose colon was perforated by the tumor and those with no perforation. The following commands will first yield KM estimates for each group separately, then plot the separate curves and place a legend in the figure. Note that in the legend command, the order of the labels corresponds to the numeric order of the values for the `perfor` variable; i.e., 0=no perforation, 1=perforation. The subcommand `lty` indicates to R what type of line should be used for each of the categories. There are a variety of choices, and the interested reader is encouraged to investigate these further. Finally, the 1000 and .9 in the `legend` command tell R where on the *x*-axis to place the left edge of the box (1000) and on the *y*-axis to place the top of the box. Again, there are many possibilities for these placement choices.

```
colon.km.fit.perforation<-survfit(colon.km.eqn~perfor ,
    data=colon_recur)
plot(colon.km.fit.perforation, xlab="Days",
ylab="Probability that cancer does not reoccur",
    main="Kaplan-Meier survival curves by perforation
status",lty=2:3)
legend(1000, .9, c("Not perforated", "Perforated"), lty = 2:3)
```

Based upon this graph, it seems clear that for this sample cancer was more likely to reoccur for individuals whose colon had been perforated. Finally, we can test whether the survival functions differ between the groups using the `survdiff` function to obtain the chi-square test results. The familiar R equation format is used here, with the survival outcome object serving as the dependent variable and perforation status as the independent variable.

```
survdiff(colon.km.eqn ~ perfor, data=colon_recur)

Call:
survdiff(formula = colon.km.eqn ~ perfor, data = colon_
recur)
```

	N	Observed	Expected	(O-E)^2/E	(O-E)^2/V
perfor=0	861	429	434.1	0.0595	2.23
perfor=1	27	17	11.9	2.1656	2.23

```
 Chisq= 2.2  on 1 degrees of freedom, p= 0.1
```

The null hypothesis for this test is that the survival rates for the two groups were the same across time. The *p*-value of 0.1 is larger than our α of 0.05, meaning that there is not sufficient evidence to conclude that survival rates for those with and without perforated colons differed across time.

Kaplan-Meier survival curves by pe

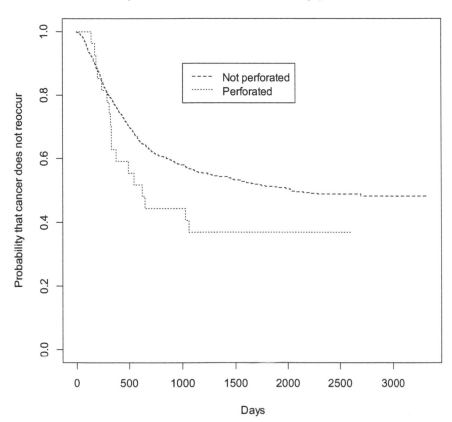

FIGURE 4.9
Kaplan–Meier survival curves by tumor perforation status

Next, let's fit a Cox proportional hazards regression model to the time until reoccurrence data. In this case, we will include the variables in Table 4.3 as predictors. We can use the coxph function to fit this model, where the output is saved in the object colon.cox.fit. The function call involves simply specifying the dependent variable object linked to the predictor variables using ~, as in the standard R regression equation format.

```
colon.cox.fit<-coxph(colon.km.eqn~sex+age+obstruct+perfor+a
dhere+nodes+differ+surg, data=colon_recur)

summary(colon.cox.fit)

Call:
coxph(formula = colon.km.eqn ~ sex + age + obstruct + perfor +
    adhere + nodes + differ + surg, data = colon_recur)
```

```
    n= 888, number of events= 446

                 coef exp(coef)  se(coef)        z Pr(>|z|)
sex        -0.154257  0.857051  0.095753  -1.611    0.1072
age        -0.003956  0.996052  0.003998  -0.989    0.3224
obstruct    0.227990  1.256073  0.118702   1.921    0.0548 .
perfor      0.191822  1.211455  0.257892   0.744    0.4570
adhere      0.236152  1.266366  0.130026   1.816    0.0693 .
nodes       0.084116  1.087755  0.009318   9.027   <2e-16 ***
differ      0.157401  1.170465  0.097500   1.614    0.1064
surg        0.265852  1.304542  0.103927   2.558    0.0105 *
---
Signif. codes:  0 '***' 0.001 '**' 0.01 '*' 0.05 '.' 0.1 ' ' 1

            exp(coef) exp(-coef) lower .95 upper .95
sex            0.8571     1.1668    0.7104     1.034
age            0.9961     1.0040    0.9883     1.004
obstruct       1.2561     0.7961    0.9954     1.585
perfor         1.2115     0.8255    0.7308     2.008
adhere         1.2664     0.7897    0.9815     1.634
nodes          1.0878     0.9193    1.0681     1.108
differ         1.1705     0.8544    0.9669     1.417
surg           1.3045     0.7666    1.0641     1.599

Concordance= 0.639  (se = 0.013 )
Likelihood ratio test= 82.89  on 8 df,    p=1e-14
Wald test             = 108    on 8 df,    p=<2e-16
Score (logrank) test = 109.9   on 8 df,    p=<2e-16
```

The variables nodes and surg were statistically significantly associated with the hazard rate, with both having positive coefficients. On the basis of these results, we would conclude that the more lymph nodes that have been detected with cancer, the greater the hazard of reoccurrence. Likewise, if the time from surgery to registration was long (versus short), the greater the hazard of reoccurrence.

In addition to the coefficient, standard error, test statistic (ratio of coefficient to standard error), and the p-value, the table also includes the value e raised to the coefficient. This statistic is the hazard ratio and reflects the relative hazard of cancer reoccurrence for individuals with different values of the independent variable. For example, the hazard ratio for surg is 1.3045, meaning that the hazard of having the cancer return is 1.3045 times higher for those whose cancer was registered a long time after its initial discovery. We can also say from this value that the hazard of cancer returning for those with long time registration is approximately 30% higher than for those with a short registration time. The hazard ratio for nodes was 1.0878, meaning that for every 1 unit increase in the number of nodes where cancer was found, the hazard of cancer returning was 1.0878 times higher. This value can also be interpreted as meaning that the risk of cancer reoccurring is approximately 9% higher for every additional lymph node in which cancer was found.

In addition to the coefficient table, R also produces an additional table including the hazard ratio, the hazard ratio of the negative coefficient value, and the upper and lower bounds of the 95% confidence interval for the hazard ratio. If the confidence interval includes 1, we conclude that there is not a relationship between the independent variable and the hazard of the event; i.e., the hazard of the event is not different for individuals with differing values on the independent variable. We can see that only nodes and surg had intervals that did not include 1, thereby reflecting the statistical significance results that we saw in the preceding table.

Finally, beneath the second table, we have several overall test statistics. The likelihood ratio, score, and Wald statistics all test the null hypothesis that all of the coefficients are 0. We see that for each, the p-value is less than 0.05 indicating that we should reject the null, and conclude that at least one of the coefficients is not 0 in the population. The concordance statistic is a measure of agreement between the estimated hazard ratio and the actual time until the event (or censoring) occurred. If for every pair of individuals in the sample the person with the longest time until cancer reoccurred also had the smallest hazard ratio, then the concordance value would be 1. Conversely, if the higher hazard value in the pair always belonged to the individual with the higher time until the event, the concordance statistic would be 0. Thus, the higher the concordance value, the more accurate is the estimated hazard ratio in capturing the time until event data. In this example, the concordance was 0.639, suggesting that there was a moderate level of agreement between the estimated hazard rate and the actual time until the reoccurrence of cancer.

Regularized Cox Regression

As with other models in this chapter, regularization methods can be applied to Cox regression for survival data in a very straightforward manner. The lasso works quite similarly to other regression models by minimizing the following function:

$$min\left\{-\sum_{(i|\delta_i=1)} ln\left[\frac{e^{\beta^T x_i}}{\sum_{j \in R_i} e^{\beta^T x_j}}\right] + \lambda \|\beta\|_1\right\}. \tag{4.21}$$

where
δ_i = Indicator whether event occurred (1) or the data were censored (0) for individual i
R_i = Risk set of individuals in the set at time t
x_i = Covariate values for individual i
β = Model coefficients
λ = Regularization parameter.

Very much as with other analyses that we've examined thus far, our primary concern in applying regularization methods such as the lasso will be the identification of the optimal value of λ. We will do so using k-fold cross-validation as we have before.

The ridge estimator takes the form:

$$min\left\{-\sum_{(i|\delta_i=1)} ln\left[\frac{e^{\beta^T x_i}}{\sum_{j\in R_i} e^{\beta^T x_j}}\right] + \lambda \|\beta\|_2^2\right\}$$ (4.22)

And finally, the elastic net is written as

$$min\left\{-\sum_{(i|\delta_i=1)} ln\left[\frac{e^{\beta^T x_i}}{\sum_{j\in R_i} e^{\beta^T x_j}}\right] + \lambda\left[\frac{1}{2}(1-\alpha)\|\beta\|_2^2 + \alpha \|\beta\|_1\right]\right\}$$ (4.23)

In order to fit the lasso estimator to the Cox model, we will use the glmnet library. First, we need to structure the data appropriately for the function.

```
Y<-cbind(time=colon_recur$time, status=colon_recur$status)

colon.predictor.list<- c("sex", "age", "obstruct",
"perfor", "adhere", "nodes", "differ", "surg")

colon.predictors <- as.matrix(colon_recur[colon.predictor.
list])
```

We put the time and status variables into the matrix Y. We then create a list of the independent variables to be included in the analysis. Finally, the matrix colon.predictors is created by subsetting the variables listed in colon.predictor.list. We are now ready to apply cross-validation to the Cox model for these variables.

```
colon.lasso.cv <- cv.glmnet(colon.predictors, Y, alpha = 1,
family = "cox", nfolds=10, type.measure="deviance")
```

We specify the independent variables, the response matrix, the value of alpha at 1 (forcing the lasso), the family as cox, 10 folds in the cross-validation, and the deviance measure as our method for identifying the optimal λ value.

Now that we've saved the lasso output in colon.lasso.cv, we can plot the log of λ, and also print the minimum λ.

```
colon.lasso.cv$lambda.min
[1] 0.0100653

plot(colon.lasso.cv)
```

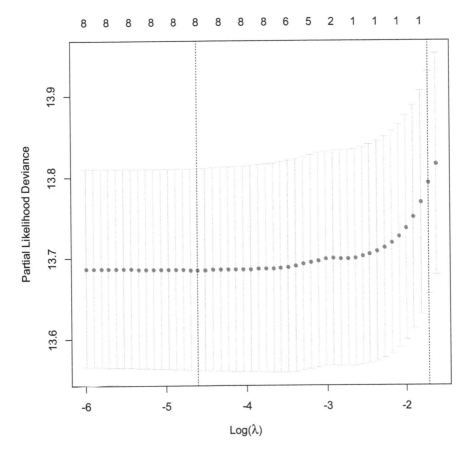

FIGURE 4.10
Log of λ by partial likelihood deviance value for lasso estimator

Now that we have identified the optimal λ, we can fit the model using the lasso estimator. The syntax is very close to what we used for the logistic regression model, with the primary difference being that the response family is cox. Note that we set lambda equal to the optimal value obtained from the cross-validation results. We can then view the coefficients using the coef function.

```
colon.lasso.optimal <- glmnet(colon.predictors, Y, alpha =
1, family = "cox", lambda = colon.lasso.cv$lambda.min)

coef(colon.lasso.optimal)
8 x 1 sparse Matrix of class "dgCMatrix"
                    s0
sex       -0.105944948
age       -0.002288979
```

```
obstruct   0.186551144
perfor     0.127396985
adhere     0.197302695
nodes      0.081547129
differ     0.120897330
surg       0.219154708
```

None of the coefficients were set to 0 by the lasso, though some are very small. In addition, each coefficient from the lasso estimator was smaller than those from the standard Cox model, which appears earlier.

We can use the `selectiveInference` library to obtain hypothesis testing results for the coefficients. As with fitting the lasso model, the use of `selectiveInference` is very similar for the Cox model as for logistic regression. Again, we don't need to estimate Sigma, so that step, which we used when modeling continuous data using lasso regression in Chapter 3, will be absent here.

```
colon.lasso.beta = as.numeric(coef(colon.lasso.optimal, colon.
predictors, Y, s=colon.lasso.cv$lambda.min, exact=TRUE))

colon.lasso.inference = fixedLassoInf(colon.predictors, colon_
recur$time,colon.lasso.beta, status=colon_recur$status,
lambda=colon.lasso.cv$lambda.min, family="cox")

colon.lasso.inference

Call:
fixedLassoInf(x = colon.predictors, y = colon_recur$time, beta =
colon.lasso.beta,
    lambda = colon.lasso.cv$lambda.min, family = "cox", status =
colon_recur$status)

Testing results at lambda = 0.010, with alpha = 0.100
```

Var	Coef	Z-score	P-value	LowConfPt	UpConfPt	LowTailArea	UpTailArea
1	-0.154	-1.611	0.268	-0.263	0.175	0.048	0.049
2	-0.004	-0.989	0.570	-0.008	0.019	0.048	0.050
3	0.228	1.921	0.121	-0.087	0.387	0.049	0.049
4	0.192	0.744	0.635	-1.593	0.513	0.050	0.049
5	0.236	1.816	0.134	-0.109	0.440	0.049	0.049
6	0.084	9.027	0.000	0.066	0.097	0.049	0.050
7	0.157	1.614	0.216	-0.141	0.280	0.050	0.050
8	0.266	2.558	0.036	0.020	0.392	0.050	0.049

Qualitatively, the results that we obtain using the lasso estimator are quite similar to those from the standard Cox regression. Namely, coefficients 6 (`nodes`) and 8 (`surg`) were found to be significantly related to the hazard of cancer reoccurring, and none of the others were.

We can fit the ridge estimator to the Cox model, by setting `alpha=0`, as follows:

```
colon.ridge.cv <- cv.glmnet(colon.predictors, Y, alpha = 0,
family = "cox", nfolds=10,type.measure="deviance")
```

```
colon.ridge.cv$lambda.min
[1] 0.221953
```

```
plot(colon.ridge.cv)
```

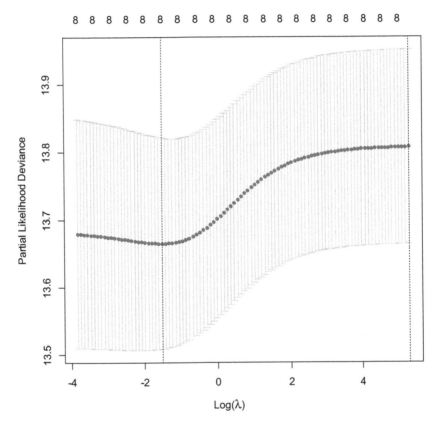

FIGURE 4.11
Log of λ by partial likelihood deviance value for Ridge estimator

With the optimal regularization parameter value in hand, we can now fit
the optimal ridge model and obtain the coefficients.

```
colon.ridge.optimal <- glmnet(colon.predictors, Y, alpha =
0, family = "cox", lambda = colon.ridge.cv$lambda.min)
```

```
coef(colon.ridge.optimal)
8 x 1 sparse Matrix of class "dgCMatrix"
                    s0
```

```
sex        -0.09645759
age        -0.00305541
obstruct    0.15654210
perfor      0.16939450
adhere      0.17756130
nodes       0.06819150
differ      0.12673696
surg        0.17735774
```

These coefficients are very similar to those that we obtained using the lasso. The inference results for the ridge estimator appears as follows:

```
colon.ridge.beta = as.numeric(coef(colon.ridge.optimal, colon.
predictors, Y, s=colon.ridge.cv$lambda.min, exact=TRUE))

colon.ridge.inference = fixedLassoInf(colon.predictors, colon_
recur$time,colon.ridge.beta, status=colon_recur$status,
lambda=colon.ridge.cv$lambda.min, family="cox")

colon.ridge.inference

Call:
fixedLassoInf(x = colon.predictors, y = colon_recur$time, beta =
colon.ridge.beta,
    lambda = colon.ridge.cv$lambda.min, family = "cox", status =
colon_recur$status)

Testing results at lambda = 0.222, with alpha = 0.100
```

Var	Coef	Z-score	P-value	LowConfPt	UpConfPt	LowTailArea	UpTailArea
1	-0.154	-1.611	0.307	-0.252	0.204	0.050	0.050
2	-0.004	-0.989	0.448	-0.009	0.013	0.050	0.050
3	0.228	1.921	0.192	-0.154	0.357	0.049	0.049
4	0.192	0.744	0.510	-1.087	0.585	0.050	0.049
5	0.236	1.816	0.178	-0.156	0.402	0.049	0.050
6	0.084	9.027	0.000	0.051	0.085	0.048	0.048
7	0.157	1.614	0.191	-0.125	0.288	0.049	0.050
8	0.266	2.558	0.090	-0.046	0.354	0.050	0.049

For the ridge estimator, only the coefficient for nodes was significantly related to the hazard rate of cancer reoccurrence. The coefficient for surg is not statistically significant using the ridge estimator, which differs from the results we obtained using the lasso and standard Cox models.

Finally, we can use the elastic net estimator with the Cox model just as we did for the other models in this chapter. The automated approach to identifying the optimal combination of α and λ based on the train function is not available for use with the Cox model, however. Therefore, we will use the same approach that was employed for the ordinal logistic regression model. Namely, we will fit a series of Cox models with glmnet, each with different values of alpha. The model which minimizes the mean cross-validation error (CVM) will be selected as optimal. For this example, we will use a small number of values for alpha, but it is important to remember that using a larger array of such values will yield a more precisely identified optimal model. The R code for fitting one of these models and then obtaining the CVM value appears as follows:

```
colon.net1.cv <- cv.glmnet(colon.predictors, Y,
alpha = 0.1, family = "cox", nfolds=10,type.
measure="deviance")

colon.net1.min.cvm<-which.min(colon.net1.cv$cvm)

colon.net1.cvm<-colon.net1.cv$cvm[colon.net1.min.cvm]

colon.net1.lambdamin<-colon.net1.cv$lambda.min
```

The first command in this sequence fits the elastic net with `alpha=0.1`, and uses the standard approach to identify the optimal λ value. We then identify the number associated with the minimum cvm and save it to the object `colon.net1.min.cvm`. Next, we save the actual cvm value itself and then identify the λ associated with it, and saving it to `colon.net1.lambdamin`. We can repeat this process for several α values, which appear in

TABLE 4.4

Optimal CVM values for each value of alpha for Cox model elastic net

Alpha	CVM
0.1	13.66509
0.2	13.66253
0.3	13.66765
0.4	13.66712
0.5	13.68015
0.6	13.6743
0.7	13.66872
0.8	13.68136
0.9	13.66565

The minimum CVM value was associated with $\alpha=0.2$. The resulting coefficients for this model, which has the optimal combination of α and λ, can then be obtained using the following commands:

```
colon.net.optimal <- glmnet(colon.predictors, Y, alpha =
0.2, family = "cox", lambda = colon.net2.cv$lambda.min)

coef(colon.net.optimal)
8 x 1 sparse Matrix of class "dgCMatrix"
```

```
                     s0
sex        -0.100330504
age        -0.002350719
obstruct    0.176540089
perfor      0.131685606
adhere      0.190001857
nodes       0.078851749
differ      0.120402634
surg        0.206316111
```

We can then obtain the inference results for this optimal elastic net Cox model using selectiveInference, just as we did for both the lasso and ridge estimators.

```
colon.net.beta = as.numeric(coef(colon.net.optimal, colon.
predictors, Y, s=colon.net2.cv$lambda.min, exact=TRUE))
```

```
colon.net.inference = fixedLassoInf(colon.predictors, colon_
recur$time,colon.net.beta, status=colon_recur$status,
lambda=colon.net2.cv$lambda.min, family="cox")
```

```
Call:
fixedLassoInf(x = colon.predictors, y = colon_recur$time, beta =
colon.net.beta,
    lambda = colon.net2.cv$lambda.min, family = "cox", status =
colon_recur$status)
```

```
Testing results at lambda = 0.046, with alpha = 0.100
```

Var	Coef	Z-score	P-value	LowConfPt	UpConfPt	LowTailArea	UpTailArea
1	-0.154	-1.611	0.293	-0.257	0.192	0.048	0.050
2	-0.004	-0.989	0.559	-0.008	0.019	0.048	0.050
3	0.228	1.921	0.143	-0.107	0.378	0.050	0.048
4	0.192	0.744	0.622	-1.532	0.523	0.050	0.049
5	0.236	1.816	0.150	-0.125	0.430	0.049	0.048
6	0.084	9.027	0.000	0.063	0.095	0.048	0.050
7	0.157	1.614	0.218	-0.143	0.280	0.049	0.049
8	0.266	2.558	0.049	0.001	0.380	0.050	0.049

On the basis of these results, we can conclude that coefficients 6 (nodes) and 8 (surg) are both significantly related to the hazard of cancer reoccurrence. As we noted earlier, having more nodes with cancerous cells and having the cancer registry late were both associated with a higher hazard for cancer to return.

Finally, we will be interested in obtaining the concordance index for the regularized models, which can be done using the Cindex function, which is part of the glmnet library. In order to do so, we first need to obtain the predicted values for the selected estimator (e.g., lasso) and then feed that to the Cindex function, along with the dependent variable, in this case Y. The predict function requires that we provide the model for which we want predictions and the set of predictor variables. We save these predictions into an object, and then pass those to the Cindex function, along with the survival dependent variable object Y.

```
#Concordance index#
lasso.pred = predict(colon.lasso.optimal, newx = colon.
predictors)

Cindex(lasso.pred, Y)
[1] 0.6423768
```

This value (0.642) is very similar to that of the standard Cox model (0.639), reflecting moderate fit of the model to the data. We can use the same command sequence to obtain concordance index values for the ridge and elastic net estimators.

```
ridge.pred = predict(colon.ridge.optimal, newx = colon.
predictors)
Cindex(ridge.pred, Y)
[1] 0.641563

net.pred = predict(colon.net.optimal, newx = colon.
predictors)
Cindex(net.pred, Y)
[1] 0.6432596
```

The concordance index results for the ridge and elastic net estimators were quite similar to that from the lasso model.

Summary

Chapter 4 serves to extend many of the principles that were introduced in Chapters 2 and 3. Specifically, we saw that the regularization techniques that were applied to models with normally distributed dependent variables, including the lasso, ridge, and elastic net, can be extended in a straightforward manner to cases where the response variable is categorical in nature. Thus, researchers who are working with dichotomous outcomes requiring the use of logistic regression can use penalized estimators for situations in which such techniques might be particularly helpful (e.g., high-dimensional data, collinearity). Similarly, logistic regression models for categorical variables with three or more outcomes that are either ordered or not can also be regularized. Finally, we saw in Chapter 4 that models involving count data and therefore requiring the use of the

Poisson distribution, as well as those for which time until even data are being modeled (Cox regression) can also be estimated using regularization techniques including the lasso, ridge, and elastic net. In short, virtually all of the approaches outlined in Chapter 3 can be easily extended to GLiMs using the `glmnet` and associated libraries in R.

In the next chapter, we will turn our attention from the univariate context, in which there is a single dependent variable, to situations in which we have multivariate data; i.e. multipole dependent variables. As we will see in Chapter 5, familiar modeling techniques such as multivariate regression, canonical correlation, and discriminant analysis can all be regularized using the principles that we outlined in Chapter 2. Furthermore, these methods can be carried out quite easily in R using readily available libraries and functions.

5

Regularization Methods for Multivariate Linear Models

In Chapter 3, we focused our attention on regularization methods for linear models. In particular, these models were univariate in nature, meaning that they have only one dependent variable. Multiple linear regression is a classic example of this model type, with one outcome variable of interest and several independent variables, sometimes referred to as predictors or covariates. As an example, in Chapter 3 we fit univariate linear models with the WAIS Verbal standard test score as the dependent variable, and several measures from the Kaufman Brief Intelligence Test (KBIT), along with some demographic factors serving as the independent variables. Using these data, we saw how to apply both the standard linear model and a variety of regularized estimators, using the R software package. Given the wide array of choices when it comes to penalized estimation, a major task for data analysts using these techniques is to determine the optimal regularization parameter values, and indeed the optimal regularization approach.

In Chapter 5, our goal is to extend the univariate models in Chapter 3 to the case when we have more than one dependent variable. In particular, we will take the univariate linear model and add dependent variables to it, yielding multivariate linear regression. In many respects, this multivariate model is quite similar to its univariate analog, both in terms of the information that we can obtain from it, and the way that we fit it using the R software package. Next, we will examine an alternative approach for estimating relationships among sets of variables using regularized canonical correlation. Unlike with multivariate regression, canonical correlation is not based on the dependent-independent variable dichotomy. Rather, the variables are divided into sets based on theory and then correlations between optimized linear combinations of these new variables are calculated. In this respect, canonical correlation can be seen as a multivariate version of correlation. After learning about the standard canonical correlation algorithm, we will then apply regularization techniques to it.

The chapter finishes with discriminant analysis, which is a related technique to canonical correlation. However, whereas with the latter we have two sets of variables for which we want to estimate relationships, in the former situation we have a single categorical variable that we want to relate to a set of (usually) continuous variables. In this sense, we can see discriminant analysis as an alternative to logistic regression, which we described in Chapter 4.

DOI: 10.1201/9780367809645-5

Discriminant analysis identifies one or more optimal linear combinations of the continuous variables that maximize the difference in the mean(s) of the resulting variates among the groups of interest. In contrast, logistic regression finds the regression coefficients that maximize the likelihood of the observed group membership for the categorical dependent variable. Although their algorithms differ, both logistic regression and discriminant analysis can be used for the prediction of group membership based on the predictors, and for identifying which of the predictors are most important for differentiating among the groups. As with multivariate regression and canonical correlation, we will first introduce the standard estimator and then discuss how it is regularized. For each of these analyses, we will demonstrate the use of R for both the standard and regularized versions.

Standard Multivariate Regression

In Chapter 3, we reviewed the basic univariate linear multiple regression model relating one or more independent variables to a single dependent or response variable. This model can be easily extended to the multivariate case in which we have more than one outcome, taking the form:

$$Y = XB + W \qquad (5.1)$$

where
   Y = Matrix of dependent variables
   X = Matrix of independent variables
   B = Matrix of regression coefficients
   W = Matrix of error terms for the two response

The assumptions underlying the multivariate regression model correspond directly to their univariate analogs, including the assumptions of normally distributed, independent, homoscedastistic error terms. In addition, the errors for the two variables are assumed to be independent.

We can fit the multivariate regression model in R using the lm function, which we first described in Chapter 3 when discussing univariate linear regression. In order to demonstrate the use of multivariate regression, we will use a dataset of Civics measures from 100 American high school students. The variables included in the dataset appear in Table 5.1.

For this example, the dependent variables are the content and skills scores, with the independent variables being the other variables in the dataset. The goal of this analysis is to understand which of the other measures are associated with the level of factual civics knowledge students have, and how well they can apply civics skills to understand current events, government

TABLE 5.1

Variables included in the Civics dataset

Variable name	Description
Gender	Self identified gender (0=Male, 1=Female)
Content	Measure of Civics content knowledge
Skills	Measure of Civics skills usage
Civics	Total Civics score (combination of Content and Skills)
Citizenship	Sense of citizenship (higher scores indicate stronger feelings of citizenship
Social	Sense of social responsibility (higher scores indicate stronger feelings of social responsibility)
Economy	Opinion regarding role of government in managing the economy (higher scores indicate greater agreement that government should manage economy)
Security	Opinion regarding role of government in ensuring personal security (higher scores indicate agreement that government should ensure personal security)
Trust	Trust in government (higher scores reflect greater trust in government)
Patriotism	Feeling of patriotism (higher scores indicate greater patriotism)
Women	Opinions regarding the place of women in society (higher scores indicate greater agreement that women should have an equal role to men)
Immigration	Attitudes towards immigration (higher scores reflect more positive attitudes to immigration to the US)
schoolA	Number of school activities; e.g., sports, music, student government
PoliticalA	Number of poitical activities; e.g., working on campaign, writing letters to government officials
Climate	Role of government in managing climate change (higher scores reflect a desire for more government involvement in countering climate change)

operations, etc. Because we are considering these two variables as a single outcome, a multivariate analysis in which they are considered together is more appropriate than two separate univariate regression models. In the latter case, any relationships between the variables (both conceptual and statistical) are ignored.

The R code to fit the standard multivariate regression model appears as follows:

```
multivariate.reg<-lm(cbind(content,
skills)~citizenship+economy+security+social+immigration+trust+
security+social+trust+patriotism+climate, data=civics)
```

Notice that the basic command structure is very similar to what we used for the univariate regression models in Chapter 3. We use the lm command with the dependent variable set on the left side of the ~ followed by the independent variables, and the definition of the dataset used in the analysis. The two dependent variables are bound together in a matrix using the cbind command and serve as the dependent entity in the equation.

Once we have fit the model to the data, we can assess the statistical significance of the relationships between each of the independent variables and the dependent variable set using the anova command.

```
anova(multivariate.reg)
Analysis of Variance Table
```

	Df	Pillai	approx F	num Df	den Df	Pr(>F)	
(Intercept)	1	0.96519	1247.62	2	90	< 2e-16	***
citizenship	1	0.03782	1.77	2	90	0.17639	
economy	1	0.01280	0.58	2	90	0.55995	
security	1	0.03890	1.82	2	90	0.16772	
social	1	0.06286	3.02	2	90	0.05385	.
immigration	1	0.01970	0.90	2	90	0.40852	
trust	1	0.01940	0.89	2	90	0.41416	
patriotism	1	0.01503	0.69	2	90	0.50586	
climate	1	0.01676	0.77	2	90	0.46748	
Residuals	91						

```
---
Signif. codes:  0 '***' 0.001 '**' 0.01 '*' 0.05 '.' 0.1 ' ' 1
```

Based on these results, there were no statistically significant relationships between the independent variables and the response set. We can obtain the partial eta-squared effect size estimates for the independent variables using the following command that is part of the heplots library in R:

```
etasq(multivariate.reg, partial=TRUE, test="Pillai")
                  eta^2
citizenship 0.007743047
economy     0.020302715
security    0.017686371
social      0.057544022
immigration 0.015122896
trust       0.013650937
patriotism  0.014959084
climate     0.016755561
```

Recall that partial eta-squared reflects the proportion of unique variance in the dependent variables that is accounted for by the independent variable. From the results above, we can see that social accounted for the most unique variance (approximately 6%), whereas citizenship accounted for the least (approximately 1%).

The assumptions underlying the multivariate regression model are essentially identical to those that we discussed in Chapter 3 for univariate regression. Therefore, we will want to use the tools that we discussed there to assess homoscedasticity, normality of the residuals, and collinearity. We will not delve into those details here, as our focus is on fitting the multivariate model. However, the assumptions were checked and all were found to have been met.

As mentioned earlier, we didn't find any statistically significant relationships between any of the independent variables and the combined responses.
If there were a statistically significant multivariate result we would then
examine the individual univariate regression models in order to determine
which of the predictors were related to each of the dependent variables. We
can use the summary command for this purpose.

```
summary(multivariate.reg)
Response content :

Call:
lm(formula = content ~ citizenship + economy + security +
    social + immigration + trust + security + social +
    trust + patriotism + climate, data = civics)

Residuals:
    Min      1Q   Median      3Q     Max
-46.272 -13.542  -0.337  14.151  53.061

Coefficients:
             Estimate Std. Error t value Pr(>|t|)
(Intercept) 106.0060    22.1923   4.777  6.8e-06 ***
citizenship  -0.8041     1.1520  -0.690   0.4869
economy      -1.8503     1.3792  -1.342   0.1831
security      1.7956     1.4510   1.237   0.2191
social       -2.9720     1.2798  -2.322   0.0225 *
immigration   0.7089     1.1389   0.622   0.5352
trust         1.3671     1.4012   0.976   0.3318
patriotism   -0.3035     1.1094  -0.274   0.7851
climate       1.5914     1.3372   1.190   0.2371
---
Signif. codes:  0 '***' 0.001 '**' 0.01 '*' 0.05 '.' 0.1 ' ' 1

Residual standard error: 22.62 on 91 degrees of freedom
Multiple R-squared:  0.1568,   Adjusted R-squared:  0.08266
F-statistic: 2.115 on 8 and 91 DF,  p-value: 0.04212


Response skills :

Call:
lm(formula = skills ~ citizenship + economy + security +
    social + immigration + trust + security + social +
    trust + patriotism + climate, data = civics)

Residuals:
    Min      1Q  Median      3Q     Max
-58.15  -14.18    2.03   15.23   46.67

Coefficients:
             Estimate Std. Error t value Pr(>|t|)
(Intercept)  96.0208    22.8437   4.203  6.14e-05 ***
citizenship  -0.9835     1.1858  -0.829   0.409
economy      -1.6547     1.4197  -1.165   0.247
```

```
security        1.6648      1.4936    1.115     0.268
social         -1.8186      1.3174   -1.380     0.171
immigration     1.3448      1.1723    1.147     0.254
trust           0.4494      1.4423    0.312     0.756
patriotism      0.6876      1.1420    0.602     0.549
climate         1.5254      1.3765    1.108     0.271
---
Signif. codes:  0 '***' 0.001 '**' 0.01 '*' 0.05 '.' 0.1 ' ' 1

Residual standard error: 23.29 on 91 degrees of freedom
Multiple R-squared:   0.1054,  Adjusted R-squared:   0.02675
F-statistic:  1.34 on 8 and 91 DF,  p-value: 0.234
```

Before walking through these results, it is very important to remember that none of the independent variables were statistically significantly related to the multivariate response. Therefore, in practice, we would not use this follow-up approach for this example. However, for pedagogical purposes, we will interpret the univariate results as if one or more of the multivariate tests were significant. It appears that social was negatively associated with the content score, such that for every 1 point increase in social, there was a 2.97-point decrease in the content test score. None of the other univariate relationships were found to be statistically significant.

Regularized Multivariate Regression

The multivariate extension to the lasso is very similar to its univariate analog, as can be seen in the lasso minimization equation as follows:

$$argmin \frac{1}{2n} \|Y - XB\|^2 + \Sigma \lambda_{jk} |\beta_{jk}|$$ (5.2)

where

λ_{jk} = Regularization parameter value for independent variable j with dependent variable k

β_{jk} = Regression coefficient relating independent variable j to dependent variable k

Thus, as in the univariate case, the lasso penalty is designed to minimize regression coefficients relating to specific independent and dependent variables. It is important to note that the regularization penalty terms are applied to each of these relationships, meaning that the relationship for one independent variable could be penalized to 0 with respect to some of the dependent variables and not for others.

The multivariate ridge (5.3) and elastic net (5.4) penalty terms are written as

$$argmin\frac{1}{2n}\|\boldsymbol{Y}-\boldsymbol{XB}\|^2 + \Sigma \lambda_{jk}\left|\beta_{jk}^2\right| \tag{5.3}$$

$$argmin\frac{1}{2n}\|\boldsymbol{Y}-\boldsymbol{XB}\|^2 + \Sigma \lambda_{jk}\left(\alpha_{jk}\left|\beta_{jk}\right|+\left(1-\alpha_{jk}\right)\beta_{jk}^2\right) \tag{5.4}$$

where

α_{jk} = Penalty mixing parameter for the combination of independent variable j and dependent variable k

As in the univariate cases, for each type of penalty, the algorithm is designed to identify the values of α_{jk} and/or λ_{jk} that minimizes the function. And, as was true in Chapter 3, the optimal solution can be identified using k-fold cross-validation and the mean squared error (MSE).

Regularized Multivariate Regression in R

We will fit the regularized multivariate regression model to the civics data using the glmnet function, just as was the case for the univariate model. First, let's see what the coefficient values look like across a variety of penalty parameter values.

```
multivariate.lasso <- glmnet(as.matrix(xvar), as.matrix(yvar),
alpha = 1, family = "mgaussian", nlambda=100)

par(mfrow=c(1,2)) #Allow for printing 2 graphs per page

plot(multivariate.lasso)
```

We can see that for both dependent variables, one independent variable had a non-zero parameter estimate for values of λ near 0. When the regularization parameter reaches approximately 3, two additional variables have non-zero parameter estimates, and an additional one is not 0 for λ of about 5.

Next, we want to apply cross-validation to determine the optimal lasso penalty parameter value using the following commands to fit the multivariate regression model and plot the MSE for 100 λ values. Note that we use 10-fold cross-validation in this example. In addition, we use the as.matrix function to ensure that our independent (xvar) and dependent (yvar) variables sets are matrices, which is required by the function. We then plot the MSE by the tuning parameter value.

```
multivariate.lasso.cv<-cv.glmnet(as.matrix(xvar),
as.matrix(yvar), family="mgaussian", type.measure="mse",
nfolds=10)

plot(multivariate.lasso.cv)
```

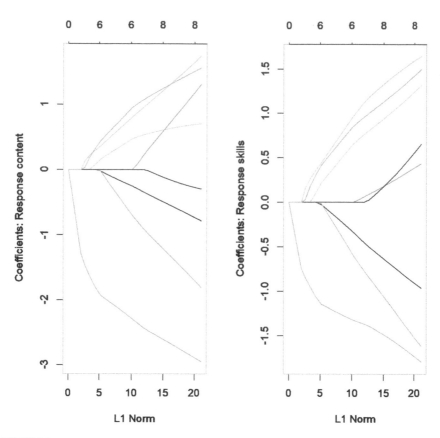

FIGURE 5.1
Coefficient values by lasso regularization parameter value

We can also print the optimal λ value.

```
multivariate.lasso.cv$lambda.min
[1] 6.013824
```

The optimal λ is 6.013824, which we can then apply in the following code. We can then use the `coef` function to see the regularized parameter estimates for each combination of the dependent and independent variables.

```
multivariate.lasso.optimal <- glmnet(as.matrix(xvar),
as.matrix(yvar), alpha = 1, family = "mgaussian",lambda =
multivariate.lasso.cv$lambda.min)

coef(multivariate.lasso.optimal)
$content
9 x 1 sparse Matrix of class "dgCMatrix"
                    s0
           112.9819107
```

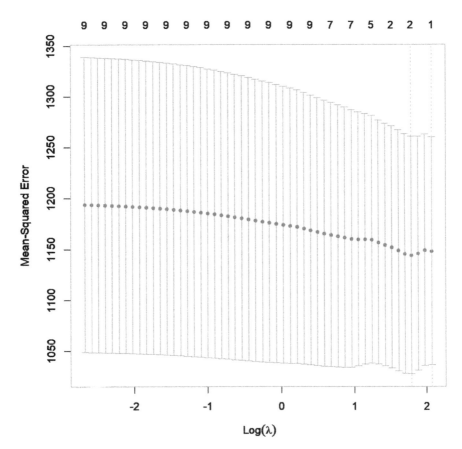

FIGURE 5.2
MSE by log of tuning parameter for lasso estimator

```
citizenship    .
economy        .
security       .
social       -0.9393304
immigration    .
trust          .
patriotism     .
climate        .

$skills
9 x 1 sparse Matrix of class "dgCMatrix"
                  s0
            115.2662359
citizenship    .
economy        .
security       .
social       -0.5456344
```

```
immigration    .
trust          .
patriotism     .
climate        .
```

Only the coefficient for `social` was not set to 0, with the relationship being negative, meaning that individuals with a higher social responsibility score had lower scores in both `content` and `skills`.

We can obtain inference information for the parameter using the `selectiveInference` library. The command sequence is carried out separately for each of the dependent variables, and is identical to that used for univariate regression in Chapter 3.

```
multivariate1.lasso.sigma<-estimateSigma(as.matrix(xvar),
civics$content, intercept=TRUE, standardize=TRUE)

multivariate2.lasso.sigma<-estimateSigma(as.matrix(xvar),
civics$skills, intercept=TRUE, standardize=TRUE)
lasso.beta = coef(multivariate.lasso.optimal,
s=multivariate.lasso.cv$lambda.min)

lasso.beta1<-lasso.beta$content[-1]

multivariate1.lasso.inference = fixedLassoInf(as.matrix(xvar),
civics$content,lasso.beta1,lambda=multivariate.lasso.
cv$lambda.min, sigma=multivariate1.lasso.sigma$sigmahat)

multivariate1.lasso.inference

Call:
fixedLassoInf(x = as.matrix(xvar), y = civics$content, beta
= lasso.beta1,
    lambda = multivariate.lasso.cv$lambda.min, sigma =
multivariate1.lasso.sigma$sigmahat)

Standard deviation of noise (specified or estimated) sigma
= 23.108

Testing results at lambda = 5.480, with alpha = 0.100

 Var   Coef Z-score P-value LowConfPt UpConfPt LowTailArea UpTailArea
   4 -3.023  -2.975   0.003    -4.697   -1.296        0.05       0.05

Note: coefficients shown are partial regression coefficients
```

The relationship between `social` and `content` was statistically significant. Next, we examine the inference results for `skills`.

```
lasso.beta2<-lasso.beta$skills[-1]

multivariate2.lasso.inference = fixedLassoInf(as.matrix(xvar),
civics$skills,lasso.beta2,lambda=multivariate.lasso.
cv$lambda.min, sigma=multivariate2.lasso.sigma$sigmahat)
```

```
multivariate2.lasso.inference

Call:
fixedLassoInf (x = as.matrix(xvar), y = civics$skills, beta
= lasso.beta2,
    lambda = multivariate.lasso.cv$lambda.min, sigma =
multivariate2.lasso.sigma$sigmahat)

Standard deviation of noise (specified or estimated) sigma
= 23.726

Testing results at lambda = 5.480, with alpha = 0.100

 Var   Coef Z-score P-value LowConfPt UpConfPt LowTailArea UpTailArea
   4 -1.756  -1.683   0.093    -3.481      0.5       0.049      0.049

Note: coefficients shown are partial regression coefficients
```

If we set $\alpha=0.05$, then we would not conclude that there was a statistically significant relationship between the social and skills scores.

Just as we were able to fit the model with the lasso penalty, we can also do so using the ridge estimator, as below. Given that we have discussed this approach in some detail previously, we will simply present the R code with the associated output, followed by a brief summary. Note that just as in the univariate case, the difference between the lasso and ridge estimator is in the setting of alpha to 1 for the former, and 0 for the latter.

```
multivariate.ridge <- glmnet(as.matrix(xvar), as.matrix(yvar),
alpha = 0, family = "mgaussian", nlambda=100)

par (mfrow=c(1,2))

plot (multivariate.ridge)
```

With the ridge penalty, we see that a number of variables have non-zero coefficients for both dependent variables with low levels of the regularization parameter.

Next, we use 10-fold cross-validation to identify the optimal λ value and plot the log of the tuning parameter values by MSE.

```
par(mfrow=c(1,1)) #Set graphic window to having 1 row and 1 column

multivariate.ridge.cv<-cv.glmnet(as.matrix(xvar), as.matrix(yvar),
family="mgaussian", type.measure="mse", nfolds=10)

plot (multivariate.ridge.cv)
multivariate.ridge.cv$lambda.min
[1] 4.992783
```

The optimal ridge penalty parameter value for this example is 4.992783. We can now use that value to fit the optimal ridge model.

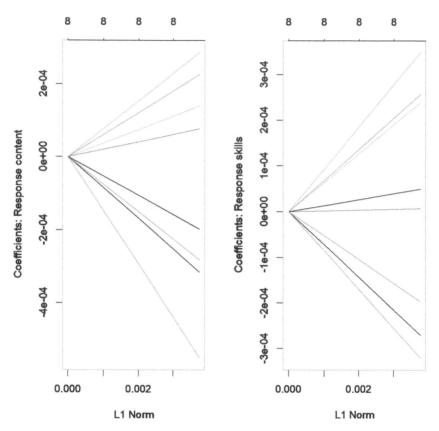

FIGURE 5.3
Coefficient values by ridge regularization parameter value

```
multivariate.ridge.optimal <- glmnet(as.matrix(xvar),
as.matrix(yvar), alpha = 0, family = "mgaussian",lambda =
multivariate.ridge.cv$lambda.min)

coef(multivariate.ridge.optimal)

$content
9 x 1 sparse Matrix of class "dgCMatrix"
                     s0
              107.6458909
citizenship   -0.2589919
economy       -0.2569918
security       0.2698882
social        -0.4814902
immigration    0.1376385
trust          0.1057759
patriotism    -0.1597815
climate        0.2255154
```

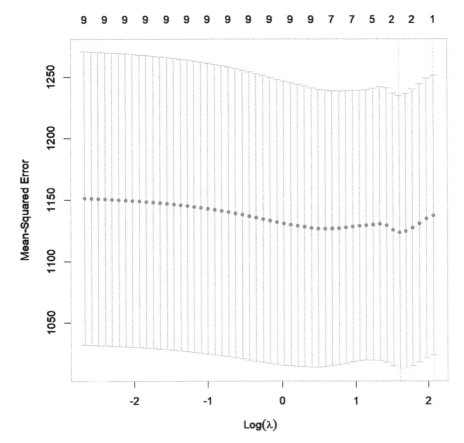

FIGURE 5.4
MSE by log of tuning parameter for ridge estimator

```
$skills
9 x 1 sparse Matrix of class "dgCMatrix"
                       s0
             108.39770268
citizenship  -0.23246932
economy      -0.19451105
security      0.30948078
social       -0.28441933
immigration   0.22338083
trust         0.02228869
patriotism    0.05335925
climate       0.23811753
```

We can see that the coefficients for trust and patriotism were shrunken to near 0, whereas those of the other variables took non-zero values.

Now that we have the ridge estimates, we can obtain the statistical significance results for the parameters using the command sequence described in
previous sections.

```
multivariate1.ridge.sigma<-estimateSigma(as.matrix(xvar),
civics$content, intercept=TRUE, standardize=TRUE)

multivariate2.ridge.sigma<-estimateSigma(as.matrix(xvar),
civics$skills, intercept=TRUE, standardize=TRUE)

ridge.beta = coef(multivariate.ridge.optimal,
s=multivariate.ridge.cv$lambda.min)

ridge.beta1<-ridge.beta$content[-1]
multivariate1.ridge.inference = fixedLassoInf(as.matrix(xvar),
civics$content,ridge.beta1,lambda=multivariate.ridge.
cv$lambda.min, sigma=multivariate1.ridge.sigma$sigmahat)
```

Following are the inference results for the content variable. It appears that
only variable 4 (social) was significantly related to content.

```
multivariate1.ridge.inference

Call:
fixedLassoInf(x = as.matrix(xvar), y = civics$content, beta
= ridge.beta1,
    lambda = multivariate.ridge.cv$lambda.min, sigma =
multivariate1.ridge.sigma$sigmahat)

Standard deviation of noise (specified or estimated) sigma = 23.237

Testing results at lambda = 4.993, with alpha = 0.100
```

Var	Coef	Z-score	P-value	LowConfPt	UpConfPt	LowTailArea	UpTailArea
1	-0.804	-0.680	0.501	-2.620	4.648	0.049	0.049
2	-1.850	-1.306	0.194	-4.177	1.836	0.049	0.049
3	1.796	1.205	0.246	-2.539	4.242	0.050	0.048
4	-2.972	-2.261	0.021	-8.758	-0.684	0.050	0.049
5	0.709	0.606	0.547	-5.263	2.470	0.050	0.050
6	1.367	0.950	0.346	-3.509	3.833	0.049	0.049
7	-0.303	-0.266	0.796	-1.616	13.010	0.049	0.050
8	1.591	1.159	0.248	-2.312	3.910	0.049	0.049

```
Note: coefficients shown are partial regression coefficients
```

Next, we can obtain inferential results for the Ridge fit of skills. The steps
match those described earlier for content.

```
ridge.beta2<-ridge.beta$skills[-1]

multivariate2.ridge.inference = fixedLassoInf(as.matrix(xvar),
civics$skills,ridge.beta2,lambda=multivariate.ridge.cv$lambda.
min, sigma=multivariate2.ridge.sigma$sigmahat)
```

```
multivariate2.ridge.inference

Call:
fixedLassoInf(x = as.matrix(xvar), y = civics$skills, beta
= ridge.beta2,
    lambda = multivariate.ridge.cv$lambda.min, sigma =
multivariate2.ridge.sigma$sigmahat)

Standard deviation of noise (specified or estimated) sigma
= 23.436

Testing results at lambda = 4.993, with alpha = 0.100
```

Var	Coef	Z-score	P-value	LowConfPt	UpConfPt	LowTailArea	UpTailArea
1	-0.983	-0.824	0.412	-2.934	3.592	0.048	0.049
2	-1.655	-1.158	0.250	-3.980	2.434	0.049	0.049
3	1.665	1.108	0.270	-2.776	4.279	0.049	0.049
4	-1.819	-1.372	0.170	-4.532	1.474	0.049	0.050
5	1.345	1.140	0.244	-2.034	4.418	0.050	0.049
6	0.449	0.310	0.765	-14.321	2.236	0.050	0.049
7	0.688	0.598	0.555	-5.312	2.404	0.050	0.050
8	1.525	1.101	0.268	-2.586	4.509	0.049	0.050

```
Note: coefficients shown are partial regression coefficients
```

None of the independent variables were associated with the skills variable, based upon these hypothesis testing results. These results are qualitatively the same as those we found when using the lasso estimator.

Standard Canonical Correlation

Canonical correlation analysis (CCA) is a statistical tool that can be used to estimate relationships between two sets of variables. In the previous section, we considered questions around relationships between sets of variables in contexts where one can be viewed as the outcome, and other set as the input, or independent variables. In contrast, with CCA the two sets are not viewed in this dependent/independent framework. Rather, our interest will be on estimating how the two sets related to one another, in much the same way we estimate relationships between individual variables using the correlation coefficient. Indeed, we can think of the difference between multivariate regression and CCA in the same way that we see the difference between univariate regression and standard correlation.

The basis of CCA centers on the creation of linear combinations for each variable to create canonical variates. These variates are calculated as weighted combinations of the variables within each set, where the weights are selected so as to maximize the correlation between the variates. These linear combinations are expressed as:

$$C_X = Z_x B_X \tag{5.5}$$
$$C_Y = Z_Y B_Y \tag{5.6}$$

where
Z_x = Matrix of standardized values of the X set
Z_Y = Matrix of standardized values of the Y set
B_X = Weights for the X set
B_Y = Weights for the Y set
C_X = Vector of canonical variates for the X set
C_Y = Vector of canonical variates for the Y set

In other words, for two sets of variables (e.g., X1-X5, Y1-Y3), the CCA algorithm identifies weights for X1-X5 in order to create a linear combination called C_X. Likewise, the algorithm will also find weights for Y1-Y3 that yields the linear combination C_Y. The weights (B_X, B_Y) within each set are determined so that the correlation between C_X and C_Y is maximized. This correlation between C_X and C_Y is the canonical correlation. The number of possible linear combinations (and thus the number of canonical correlations) is equal to the number of variables in the smaller set. Thus, in this example there are 3 possible canonical variate sets, given the 3 variables in set Y. Within a variable set (X or Y) the linear combinations are orthogonal to one another. Finally, the weights are determined such that the first canonical correlation will be the largest, the second correlation is the second largest, and so on.

In addition to the canonical correlation(s) between the two variable sets, CCA also yields additional information that can be used to more fully understand the relationships between the two sets. There is a statistical test for the null hypothesis that each canonical correlation, and all those after it, are equal to 0. When this F-test yields a statistically significant result, we conclude that the relationship between the two variable sets is not 0; i.e., the sets are related to one another. In addition, the squared value of each canonical correlation reflects the proportion of shared variance between the two variable sets. We can calculate the correlation between each variable and each canonical variate in order to ascertain the relative importance of the variables in determining the canonical variate for its set. These statistics are typically referred to as structure coefficients, or CCA loadings. Another commonly used statistical tool often used to characterize the CCA solution is the proportion of variance in the variables for one set that are accounted for by the canonical variates for the other set. This characterization is known as redundancy analysis. It is also possible to examine the actual canonical weights associated with each observed variable. However, these weights are not bounded, unlike the structure coefficients, and thus are not as useful for fully understanding the canonical variates and correlations between them.

Standard CCA in R

We will take the civics education data that was described earlier, and use CCA to gain some insights into relationships between the content and skills test scores and measures of activism in schools and political organizations. These latter two measures are counts of the number of different activities that students engaged in at school (e.g., student government, clubs, drama, music, athletics) and in the broader political culture (e.g., volunteering on campaigns, contacting political representatives). The hypothesis of interest is that students who have higher scores on the content and skills tests (set1) will also be more engaged both at school and in the broader political community (set2). We do not have a sense that there is a dependent/independent variable structure, and thus are interested in using correlation rather than regression.

First, we need to create the two sets of variables, which will be placed into matrices.

```
set1 <- civics[,c("patriotism","citizenship","trust")]
set2 <- civics[,c("schoolA","politicalA")]
```

Now we can use the cca function from the yacca library in R to fit CCA for these sets, saving the output in a file called cca.fit.

```
cca.fit <- cca(set1, set2)
```

Our first step in assessing the results is to examine the correlations themselves, along with the hypothesis tests for whether they are equal to 0.

```
F.test.cca(cca.fit)

F Test for Canonical Correlations (Rao's F Approximation)

          Corr         F  Num df  Den df     Pr(>F)
CV 1  0.61401  9.77534  6.00000     190  2.203e-09 ***
CV 2  0.25056       NA  2.00000      NA         NA
---
Signif. codes:  0 '***' 0.001 '**' 0.01 '*' 0.05 '.' 0.1 ' ' 1
```

The null hypothesis being tested by the F statistics is that the current correlation and all succeeding ones are equal to 0. The first canonical correlation value is 0.61, and the null hypothesis is rejected. For the second correlation, there is not a hypothesis test because there are no succeeding correlations; i.e., the second correlation is the final one in the set. Based on these results, we conclude that there is at least one statistically significant relationship between the two sets, and the strongest of these is 0.61, which falls into the large category based on commonly used guidelines for interpreting correlations as an effect size (Cohen, 1988). The second correlation falls into the small range and is quite a bit lower than the first one.

Given that we have a significant and large correlation, we will want to examine the structural correlations for each set in order to gain insights into the nature of these relationships. We can obtain these values by simply typing in the name of the output object.

```
cca.fit
Canonical Correlation Analysis
```

Here are the two canonical correlation coefficients.

```
Canonical Correlations:
     CV 1       CV 2
0.6140113 0.2505580
```

The canonical coefficients (B_X, B_Y) appear next. Because these values are not bounded in the way that correlations are, we will be more interested in the structural correlations, which appear after the coefficients themselves.

```
X Coefficients:
                    CV 1              CV 2
patriotism    0.27441859  -0.342981847
citizenship  -0.30431578  -0.247358102
trust        -0.05806433  -0.008461855

Y Coefficients:
                 CV 1          CV 2
schoolA     0.01949261  -0.59923205
politicalA -0.54280163  -0.09862906
```

The structural correlation for each set appears next. Recall that these are the correlation coefficients between each observed variable and each canonical variable. Larger values indicate that the observed variable contributes relatively more to the canonical variable. A common rule of thumb is that structural correlations greater than 0.3 are indicative of an "important" contribution from an observed variable. It is also important to keep in mind that variables with larger values contribute more to the canonical variable, above and beyond whether it contributes at all (i.e., has a correlation of 0.3 or more).

```
Structural Correlations (Loadings) - X Vars:
                    CV 1          CV 2
patriotism    0.6095166  -0.7912296
citizenship  -0.7747141  -0.6263463
trust        -0.3964431  -0.2509463

Structural Correlations (Loadings) - Y Vars:
                 CV 1          CV 2
schoolA     0.1787764  -0.98388973
politicalA -0.9994713  -0.03251212
```

We will initially focus on the first column of structural correlations, which pertain to the two canonical variables associated with the first canonical correlations. For the first set of variables, citizenship had the strongest relationship with the canonical variable, followed by patriotism. Trust also played a role in determining the canonical variable, though not as much as the other two. The fact that patriotism had a positive correlation with the canonical variable whereas the other two variables had negative correlations indicates that individuals with higher patriotism scores also had higher canonical variate scores, but those with higher citizenship and trust scores had lower canonical variate scores. For the second variable set, only political activities were associated with the canonical variable, and this relationship was inverse.

We can combine these results with the canonical correlation to reach some overall conclusions about how the variables in the two sets are related to one another. First, recall that the first canonical correlation was large and positive. Thus, individuals with larger values for the canonical variable in set1 also had larger values for set2. Next, we note that the relationships between political activism and its canonical variable have the same sign as citizenship and trust do with their canonical variable. This means that students who had higher political activism scores also had higher citizenship and trust scores, but that those with higher patriotism scores had lower political activism scores. In summary, the first canonical correlation was assessing the relationship between political activism primarily, with scores on citizenship, patriotism, and to a lesser degree trust in institutions. Individuals who were more politically active also had a higher sense of citizenship, and more trust in institutions, but less of a sense of patriotism about their country.

The second canonical correlation is much weaker than the first and appears to be assessing the relationship between patriotism and involvement in school activities. No other observed variable exhibited an absolute value structural correlation greater than 0.3. Based on these results, we would conclude that students who had higher scores on the patriotism scale were also more involved in school activities.

Finally, the cca function provides the redundancy statistics for each variable set. Redundancy reflects the proportion of variance in one set accounted for by the other set.

```
Aggregate Redundancy Coefficients (Total Variance
Explained):
X | Y: 0.1644921
Y | X: 0.2247501
```

From these results, we can see that the Y variables (school and political activities) accounted for 16.4% of the variance in the X variables (patriotism, citizenship, and trust). The X variables accounted for 22.5% of the variance in the Y set. We should note here that there are a number of other statistics available from the cca functions, and the interested reader is encouraged to view the online documentation to learn more about these.

Regularized Canonical Correlation

Regularized CCA (RCCA) has the same general goal as standard CCA, maximization of the correlation between the canonical variates associated with the separate variable sets. In addition to considering this correlation maximization goal when determining the canonical weights, the RCCA algorithm also has built in a penalty (e.g., lasso, ridge) on the weights (B_X, B_Y). The effect of applying the sparse penalty to the CCA maximization function is to reduce the magnitude of the weights, driving some to 0. As a result, we would expect fewer variables in each set will be associated with the resulting canonical correlations. As with other applications of these regularization methods, a key aspect of using RCCA is the determination of the optimal value of the tuning parameter, λ. In the next section, we will describe how these values can be determined, in this using a permutation-based approach. As with standard CCA, when using RCCA we will be primarily focused on the values of the correlations themselves, as well as the relationships between the individual variables and the canonical variates through the structure coefficients.

RCCA can be carried out using the PMA library in R. Our first goal in applying the lasso penalty to the CCA problem is to identify the optimal regularization parameter values for each of the variable sets, which we can do using the CCA.permute function. This approach to determining the optimal values involves the following steps:

1. The rows of the first variable set are randomly permuted.
2. Sparse CCA is applied to each of these permuted datasets and the canonical correlation values are retained.
3. Repeat steps 1–2 m (e.g., 1000) times to obtain a distribution of the canonical correlations (cca_i).
4. Apply CCA to the original data and retain the canonical correlation values (cca_o).
5. Apply Fisher's transformation to convert the canonical correlations to an approximate standard normal distribution.
6. Calculate $z = \dfrac{cca_o - \bar{x}_{cca_i}}{s_{cca_i}}$, where $\bar{x}_{cca_i}$ = Mean of permuted correlations and s_{cca_i} = Standard deviation of permuted correlations.
7. The maximum z statistic is associated with the optimal tuning parameter.

In order to employ the permutation procedure for identifying the optimal regularization parameter for each set, we need to identify the two variable sets, which are matrices just as was true for the cca function. We then indicate that for each variable we will use the standard lasso penalty. The alternative

is to use `ordered`, which would lead to the use of the fused lasso penalty. We
then define the number of permutations, with the default being 25.

```
cca.lasso.perm.out <- CCA.permute(set1, set2,typex="standar
d",typez="standard",nperms=1000)
```

We can then view the cross-validation results.

```
print(cca.lasso.perm.out)
```

```
Call: CCA.permute(x = set1, z = set2, typex = "standard",
typez = "standard",
    nperms = 1000)
```

```
Type of x:   standard
Type of z:   standard
```

	X Penalty	Z Penalty	Z-Stat	P-Value	Cors	Cors Perm	FT(Cors)	FT(Cors Perm)	# U's Non-Zero
1	0.100	0.100	3.339	0	0.481	0.155	0.524	0.156	1
2	0.167	0.167	3.339	0	0.481	0.155	0.524	0.156	1
3	0.233	0.233	3.339	0	0.481	0.155	0.524	0.156	1
4	0.300	0.300	3.339	0	0.481	0.155	0.524	0.156	1
5	0.367	0.367	3.339	0	0.481	0.155	0.524	0.156	1
6	0.433	0.433	3.339	0	0.481	0.155	0.524	0.156	1
7	0.500	0.500	3.339	0	0.481	0.155	0.524	0.156	1
8	0.567	0.567	3.339	0	0.481	0.155	0.524	0.156	1
9	0.633	0.633	3.655	0	0.517	0.162	0.572	0.164	2
10	0.700	0.700	4.045	0	0.558	0.170	0.630	0.172	2

```
     # Vs Non-Zero
1
2              1
3              1
4              1
5              1
6              1
7              1
8              1
9              1
10             1
Best L1 bound for x:   0.7
Best L1 bound for z:   0.7
```

The first two columns include the tuning parameter (parameter value) for
each variable set. Next, the z value from step 6 of the algorithm is described
earlier, along with the *p*-value associated with that z. The CCA and the permuted CCA values appear in the next two columns, followed by Fisher's z
transformation values for the CCA and permuted CCAs. Finally, the number of non-zero canonical weights for each variable set appears in the final
two columns. The best lasso penalty parameter is presented below this table.
Note that for each variable set an optimal value is selected, which for this

example happens to be 0.7 for each. Although the penalty values are the same for each set in this example, such will not always be the case.

Once we have determined the optimal regularization parameters for each set, we can then apply these values in estimating the regularized CCA values using the CCA function. We will once again need to define the matrices containing the two variable sets, as well as the type of penalty to use, just as with CCA.permute. We also provide the penalty for each set, which we do with the penaltyx and penaltyz subcommands. Here we refer to the optimal penalty values as determined by the CCA permutation procedure. We then indicate that we want to retain 2 canonical correlations.

```
cca.lasso.out <- CCA(xvar,yvar,typex="standard",typez="stand
ard",penaltyx=cca.lasso.perm.out$bestpenaltyx, penaltyz=cca.
lasso.perm.out$bestpenaltyz, K=2, v=cca.lasso.perm.out$inits)
```

We can print the output with all of the available parameter estimates, including the variable weights, by using the verbose=TRUE option.

```
print(cca.lasso.out,verbose=TRUE)
Call: CCA(x = xvar, z = yvar, typex = "standard", typez =
"standard",
penaltyx = cca.lasso.perm.out$bestpenaltyx, penaltyz =
cca.lasso.perm.out$bestpenaltyz, K = 2, v=cca.lasso.perm.
out$inits)

Num non-zeros u's:  1 1
Num non-zeros v's:  2 3
Type of x:   standard
Type of z:   standard
Penalty for x: L1 bound is  0.7
Penalty for z: L1 bound is  0.7
Cor(Xu,Zv):   0.5577731 0.2695679

 Component  1 :

  Row Feature Name Row Feature Weight
1                2                  1

   Column Feature Name Column Feature Weight
1                    1                 -0.242
2                    2                  0.970

 Component  2 :

  Row Feature Name Row Feature Weight
1                1                  1

   Column Feature Name Column Feature Weight
1                    1                  0.976
2                    2                  0.219
3                    3                  0.018
```

The first canonical correlation value was 0.56, with the second being 0.27. Thus, the first regularized correlation was somewhat smaller than for the standard approach, whereas the second one was similar in value to that from the standard CCA. For the first canonical correlation, the second variable in set2 (politicalA) had a non-zero weight, which was estimated as 1. For set 1, the first two variables (patriotism and citizenship) had non-zero coefficient values, and the signs for these variables are opposite, as was the case with the standard CCA. The weight for trust was regularized to 0, meaning that it didn't contribute to the first canonical correlation at all, which differs from the standard analysis, where it had a weight of −0.06. For the second canonical correlation, only the first variable (schoolA) in set2 had a non-zero weight, which was 1. For the second set, all three variables had non-zero weights, with patriotism having the largest, and trust having the smallest. These results were generally similar to those for the standard CCA approach.

Standard Linear Discriminant Analysis

A common research question in a variety of fields involves the prediction of group membership using a set of variables, and differentiating individuals in these groups. For example, the researcher who collected the eating disorder data that we discussed in Chapter 4 may be interested in developing a prediction equation for identifying individuals who are likely to suffer from anorexia, versus other eating disorders. In addition, they would also like to know which of the measured variables in Table 4.1 most differentiate the individuals in the two diagnostic groups. In Chapter 4 we described the logistic regression model and demonstrated how it could be used to identify statistically significant predictors of group membership. An alternative analytic framework is discriminant analysis (DA), which is closely associated with CCA. Whereas with CCA the canonical weights are identified to maximize the correlation between the resulting canonical variates, the goal in DA is to identify weights that maximize the difference on the mean of the resulting discriminant variates, which are linear combinations of the predictor variables. The number of possible discriminant functions is the minimum of the number of groups minus 1 and the number of predictor variables.

Discriminant function k for individual i is defined as:

$$D_{ki} = w_o + w_1 z_{1i} + w_1 z_{1i} \ldots w_j z_{ji} \tag{5.7}$$

where

w_o = Intercept
z_{ji} = Standardized score for individual i on variable j
w_j = Discriminant function weight for variable j

Thus, the values of w_j and w_o are determined by the DA algorithm such that the group means of D_{ki} are maximally different among the groups. Each D_{ki} is orthogonal to the others. Conceptually, this means that they represent different ways in which the groups can be differentiated from one another.

In the context of explanation (identifying which variables are most important in terms of differentiating the groups), we can calculate the correlations between each of the predictors and D_{ki}. These values are known as structure coefficients, just as they were with CCA. This approach is very similar to the way in which we interpret CCA, as described earlier. And as with CCA, a commonly used cut-value for identifying important variables is an absolute value of the structure coefficients of 0.3 or larger. In other words, variables that have an absolute value of the structure coefficient of 0.3 or larger are considered to contribute to group differences, with larger coefficients contributing more to the difference. Finally, predictions of group membership for individuals can be obtained by applying equation (5.X) to an individual's scores on the predictor variables.

In order to demonstrate the fitting of standard DA in R with the `DiscriMiner` library, we will use the eating disorder data that is described in Chapter 4, and which variables appear in Table 4.1. First, we need to read the data using the `foreign` library and create a factor version of the outcome variable `diag2`.

```
library(foreign)
library(DiscriMiner)

eat_disorder_data<-read.spss("//Users/holmesfinch/
Documents/research/regularization book/chapter 4 logistic
regression data.sav", to.data.frame=TRUE, use.value.
labels=FALSE)

eat_disorder_data$diag2.f<-as.factor(eat_disorder_
data$diag2)
```

Next, we fit the standard DA model to the data with diagnosis (Anorexia or not) and measures of behavior and relationships, which are described in Chapter 4. The function call for desDA requires the set of predictors to be listed first, followed by the categorical outcome variable, which must be a factor.

```
#STANDARD DISCRIMINANT ANALYSIS#
eat_disorder.desDA<-desDA(eat_disorder_data[,3:17],eat_
disorder_data$diag2.f)
```

We can obtain all of the output produced by this function by simply typing the name of the output object.

```
eat_disorder.desDA
```

The discriminant structure coefficients (correlation between the individual variables and the discriminant function score) are contained in the `discor` object within the output object. As we noted earlier, these values are particularly useful for interpretive purposes. A commonly used cut-value for identifying important variables in terms of differentiating the groups is 0.3 (Tabachnick & Fidell, 2019). Therefore, on the basis of the values below we would conclude that purging, binging, and family relationships best differentiated individuals diagnosed with anorexia from those diagnosed with another eating disorder. In addition, the difference in signs between family relationships vis-a-vis binging and purging indicates that the group with higher scores on the latter two had lower scores on family relationships. We would also conclude that no other variables contribute to the group differences.

```
eat_disorder.desDA$discor
                DF1
fast    -0.26896367
binge    0.52324257
vomit    0.25990032
purge    0.72193665
hyper    0.14526806
fami    -0.48249290
eman    -0.26724303
frie    -0.14748392
school  -0.15272531
satt    -0.19879950
sbeh    -0.03133854
mood     0.15700097
preo    -0.19119999
body    -0.03215740
time     0.00367355
```

Next, we have the power results for each of the variables. These results provide information regarding the change in group differentiation, with statistically significant results indicating that the individual predictor was associated with group differentiation. On the basis of these results, we would conclude that fasting, binging, vomiting, purging, family relations, emancipation, sexual attitude, and preoccupation with weight all had two significantly different means between the groups.

```
Eat_disorder.desDA$power
          cor_ratio wilks_lamb  F_statistic      p_values
fast    2.785866e-02  0.9721413   6.161257247  1.382347e-02
binge   1.054336e-01  0.8945664  25.339911156  1.015058e-06
vomit   2.601277e-02  0.9739872   5.742114714  1.741918e-02
purge   2.007110e-01  0.7992890  53.989072503  4.139578e-12
hyper   8.126683e-03  0.9918733   1.761552507  1.858387e-01
fami    8.965095e-02  0.9103490  21.173147955  7.168251e-06
eman    2.750337e-02  0.9724966   6.080456523  1.445124e-02
```

```
frie     8.376496e-03   0.9916235   1.816159749 1.791894e-01
school   8.982455e-03   0.9910175   1.948732284 1.641633e-01
satt     1.521961e-02   0.9847804   3.322788403 6.971475e-02
sbeh     3.782079e-04   0.9996218   0.081345463 7.757569e-01
mood     9.492437e-03   0.9905076   2.060432466 1.526205e-01
preo     1.407825e-02   0.9859217   3.070045329 8.117288e-02
body     3.982308e-04   0.9996018   0.085653733 7.700593e-01
time     5.196907e-06   0.9999948   0.001117341 9.733654e-01
```

Finally, the discriminant weights (w_j) are also displayed. As we noted earlier, the structure coefficients are more useful for interpreting the nature of group differences. However, it is useful to have the discriminant weights if we would like to predict group membership for new cases.

```
eat_disorder.desDA$discrivar
                    DF1
constant  -0.96470340
fast      -0.30650459
binge      0.02584326
vomit     -0.04864329
purge      0.85127648
hyper      0.21019292
fami      -0.55949710
eman      -0.30201442
frie      -0.01564093
school    -0.08053796
satt      -0.10615459
sbeh       0.16417041
mood       0.36420531
preo      -0.68231575
body       0.23038432
time       0.22638867
```

Regularized Linear Discriminant Analysis

As with CCA, the lasso and ridge penalties can be applied to DA. These regularization techniques can be particularly useful when we have a large number of variables and/or a relatively small sample size. These algorithms incorporate a penalty for values of w_j that are not 0, thereby tending to drive them down in value in much the same manner as with regression models. In other respects, regularized DA is quite similar to standard DA, including in how we will interpret the results. We will use the `penalizedDA` library in order to apply the lasso and ridge estimators to the DA problem.

In order to fit penalized estimators in the context of DA, we will need to install and then load the `penalizedDA` library.

```
library(penalizedLDA)
```

The `penalizedDA` function requires that the groups be numbered sequentially starting with 1 (e.g., 1, 2, 3. . .). Currently, the anorexia diagnosis variable (`diag2`) is coded as 1 (anorexia) and 0 (other). Therefore, we will first recode the 0 to be 2 using the `ifelse` command and put them in a new variable called `diag2b` in the `eat _ disorder _ data` object.

```
eat_disorder_data$diag2b<-ifelse(eat_disorder_
data$diag2==0,2,1)
```

Our first task in fitting the lasso DA estimator to the eating disorder data problem as discussed earlier is to identify the optimal value of λ. We can do this with a cross-validation approach just as we demonstrated with regression models in Chapters 3 and 4. In this case, we will save the results from the `penalizedLDA.cv` function in the object `eat _ disorder.lasso.da.cv`. The function call requires that the predictors be in the form of the matrix, which we can ensure using the `as.matrix` function. Likewise, the response variable should be a vector object. In this example, we will use 30 cross-validation folds.

```
eat_disorder.lasso.da.cv <- PenalizedLDA.cv(as.
matrix(eat_disorder_data[,3:17]),as.vector(eat_disorder_
data$diag2b),nfold=30)
```

Next, we can fit the penalized LDA based on the optimal value of lambda.

```
eat_disorder.lasso.da <- PenalizedLDA(eat_disorder_
data[,3:17],as.vector(eat_disorder_data$diag2b),lambda=eat_
disorder.lasso.da.cv$bestlambda,K=1)
```

We can see the weights applied to each variable.

```
eat_disorder.lasso.da$discrim
              [,1]
 [1,] -0.215529929
 [2,]  0.437094617
 [3,]  0.208069703
 [4,]  0.638007594
 [5,]  0.115244633
 [6,] -0.399544936
 [7,] -0.214112000
 [8,] -0.117017263
 [9,] -0.121212945
```

```
[10,]  -0.158279350
[11,]  -0.024765071
[12,]   0.124638469
[13,]  -0.152140664
[14,]  -0.025412422
[15,]   0.002902457
```

We can obtain the correlation between the discriminant function value (xproj) and the individual variables used in the regularized DA. These correspond to the structure coefficients that we get from standard discriminant analysis.

```
cor(eat_disorder.lasso.da$xproj,eat_disorder_data[,3:17])
        fast      binge      vomit      purge      hyper
    fami          eman          frie
[1,] -0.2239257 0.6769887 0.3846361 0.7114347 0.0511126
  -0.5369755 -0.3432443 -0.1849432
        school      satt       sbeh       mood          preo
    body        time
[1,] -0.2640767 -0.2984295 -0.122412 0.1087934 -0.07162538
  -0.06780004 -0.142686
```

As with the standard DA, we will use the 0.3 cut-value to identify important variables with respect to differentiating individuals diagnosed with anorexia versus those with other diagnoses. These results reveal that in terms of behaviors, purging, binging, and vomiting were most important in separating the two groups. In addition, the family relations, emancipation, and (just

TABLE 5.2

Structure coefficients for standard DA and lasso DA

Variable	Standard DA	Lasso DA
Fast (Fasting)	−0.27	−0.22
Binge (Binging)	0.52	0.68
Vomit (Vomiting)	0.26	0.39
Purge (Purging)	0.72	0.71
Hyper (Hyperactivity)	−0.15	0.05
Family (Family relations)	−0.48	−0.54
Emancipation (Emancipation from family)	−0.27	−0.34
Friends (Number of friends)	−0.15	−0.19
School (School/employment record)	−0.15	−0.26
Satt (Sexual attitude)	−0.20	−0.30
Sbeh (Sexual behavior)	−0.03	−0.12
Mood (Mood state)	0.16	0.11
Preo (Preoccupation with weight)	−0.19	−0.07
Body (Body perception)	−0.03	−0.07
Time (Time spent thinking about weight)	0.004	−0.14

barely) the `satt` scores were also useful for group differentiation. The difference in signs between these latter three variables and binging, purging, and vomiting indicates that the group with higher scores on the latter three had lower scores on the former two.

Finally, in order to compare the results of the standard and lasso DA results, we can view the two sets of structure coefficients together in the same table.

When comparing the results, we see that the general pattern with respect to which variables are important was similar between the two estimators. The two exceptions to this similarity were the structure coefficients for emancipation and sexual attitude, both of which exceeded the 0.3 threshold for the lasso estimator, whereas for standard DA they did not. For the other variables, qualitatively the results were similar across the two estimators.

Summary

Our focus in Chapter 5 was on fitting regularized multivariate models to data. These models each have multiple dependent variables, which differentiates them from the univariate models that were the topic of Chapter 3. However, despite this fundamental difference in structure, the underlying concepts that we discussed in Chapters 2 and 3 applied here in Chapter 5 as well. In addition, employing regularization techniques using R also takes much the same form as what we saw in the earlier chapters as well. In short, although the models to which we apply them may differ, the basic principles of regularized estimation remained in place for the multivariate models. Furthermore, applying them in R involved many of the same command structures that were used for univariate linear models (Chapter 3) and GLiMs (Chapter 4).

In Chapter 6, we will examine another family of multivariate techniques, cluster analysis. These methods are useful for identifying potential underlying subgroups within the population, based on scores for multiple observed variables. We will describe two broad approaches to clustering, one in which we begin the analysis with hypotheses regarding the number of groups in the population (k-means) and the other where we simply agglomerate the observations from many clusters to just one and then use various statistical methods to identify the optimal solution. These methods can be regularized so that only those variables that truly contribute to identifying the subgroups are retained in the analysis. And again, although the overall framework for clustering differs from the other multivariate methods we consider in this book (thereby necessitating its own chapter), the basic principles of regularization that we have discussed heretofore will continue to prove useful in Chapter 6.

6

Regularization Methods for Cluster Analysis and Principal Components Analysis

With the advent of interest in machine learning algorithms, one of the increasingly popular statistical methods in recent years has been cluster analysis (CA). Much as with factor analysis, which we cover in Chapter 7, CA refers to a wide array of statistical methods, rather than a single statistical model. Although they differ in technical aspects, all CA techniques have as a common purpose the identification of subgroups within a broader population based upon scores on a set of variables. For example, a psychologist may be interested in assessing whether there are different typologies of cognitive functioning, based upon a set of scores on measures of executive functioning and cognition. Biologists might want to assess a hypothesis about differing species of flies based on clusters of gene expressions. In each of these cases, researchers believe that there may be subgroups within the population that share common patterns on the observed variables

High-dimensional data is an issue that researchers often face in conducting CA. They may be able to obtain many measured scores for a relatively small number of individual observations. The purpose of this chapter is to describe the application of regularization techniques, including the lasso and Ridge estimators, to the problem of CA. Specifically, we will see how to apply these penalized estimators to the two primary clustering frameworks, K-means and hierarchical clustering. In most respects, the principles that we discussed in Chapter 2 will be applied here just as they were to the models described in Chapters 3 through 5. We will highlight where there are unique aspects to regularized CA. As in the other chapters, our focus will be on the application of these methods using the R software package. After describing the basic principles underlying each of K-means and hierarchical clustering, we will turn our attention to their applications, followed by a description of the regularized estimators.

K-means

K-means clustering is one of the most popular applications of CA in practice. It has been used in a wide array of applications across fields as diverse as

DOI: 10.1201/9780367809645-6

psychology, education, medicine, marketing, and ecology. Researchers using this method begin the analysis by hypothesizing the number of clusters that they believe may be present in the population, based on a set of measured variables, x. The optimal solution for a given sample and a prespecified number of k clusters is determined by minimizing the within-cluster sum of squares:

$$SS_{wpk} = \sum_{r=1}^{K} \sum_{i \in K_r} \left(x_{ij} - \overline{x}_{jr} \right)^2 \qquad (6.1)$$

where

x_{ij} = Value of variable j for individual i

$\overline{x}_{jr}$ = Mean of variable j for cluster r

Thus, for each cluster r within the full set of k clusters, the sum of squares is calculated for each of the p variables. These individual variable sums of squares are then summed across all variables used in the analysis and all observations in the sample to obtain a total sum of squares for the entire cluster solution.

$$SS_w = \sum_{j=1}^{p} SS_{wpk} \qquad (6.2)$$

To begin the K-means algorithm, a set of k initial cluster centroids are selected randomly from the sample, where k is defined by the data analyst. Each member of the sample is then assigned to the cluster for which its value from equation (6.2) is minimized. The cluster centroids are updated, the sums of squares in equations (6.1) and (6.2) are recalculated, and individuals in the sample are once again placed in the cluster leading to the overall smallest sum of squares. This series of steps is repeated until no switching of cluster membership by individual observations results in a lower SS_w, at which point the optimal K-means cluster solution is achieved.

Determining the Number of Clusters to Retain

There are a number of approaches for determining the optimal value for k, the number of clusters to retain. It is important to note here, however, that there is no single definitive technique for this purpose. Rather, the researcher will apply several of these methods in their analysis and then make a decision regarding the number of clusters to retain by considering all of these. The factoextra R library provides multiple tools for helping the data analyst determine the number of clusters to retain. One useful statistic for this purpose is the within-cluster sum of squares (WSS) for a given value of k. Larger values of WSS indicate greater spread within the clusters; i.e., less within-cluster cohesion. Therefore, we may decide that the best solution is the one where the decline in WSS becomes less marked as the number of clusters increases. Using the `fviz _ nbclust` function, we can obtain a graphical representation of the WSS by the number of clusters based on the K-means algorithm. Essentially, cluster solutions from 1 through 10 are fit to the data, and for each, the WSS is calculated. The full set is then graphed, and we look for the point where the line first makes a bend. To use the function, we simply must provide the dataset, the clustering algorithm, and the fit method. The results appear in Figure 6.1.

```
fviz_nbclust(civics.cluster.data, kmeans, method="wss")
```

It would appear in this example that the bend may occur at 5 clusters. Clearly, interpretation of this plot is subjective in nature.

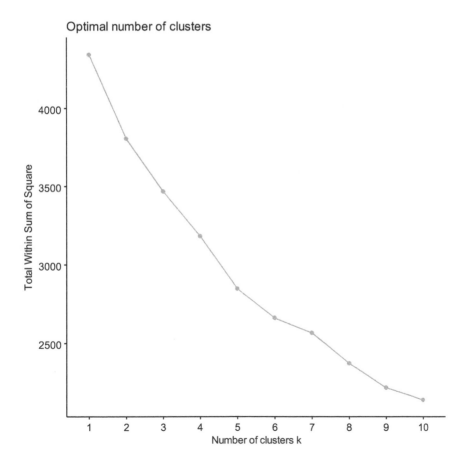

FIGURE 6.1
WSS by number of clusters for the k-means solution

A second approach for determining the number of clusters to retain is based on the silhouette statistic (Rousseeuw, 1987), which reflects the distance of individuals in one cluster from points in the other clusters. More specifically, for each data point, i, the silhouette statistic s_i is calculated using the following steps:

1. For each i, calculate the mean distance (e.g., Euclidean) $\bar{a}_i$ to all other points for the cluster to which individual i belongs.
2. Calculate the mean distance $\bar{d}_{ic}$ between individual i and all members of each cluster c, to which i doesn't belong; e.g., if individual i belongs to cluster 1, we will calculate $\bar{d}_{ic}$ separately for clusters 2, 3, ..., K.
3. Define $\bar{b}_i = min(\bar{d}_{ic})$; this is the distance from individual i to the nearest neighboring cluster.
4. Calculate the silhouette value for observation i:

$$S_i = \frac{\bar{b}_i - \bar{a}_i}{max(\bar{a}_i, \bar{b}_i)} \tag{6.3}$$

Observations with a large value of S_i are considered to be well clustered, whereas small values indicate that the observation lies near the border between two clusters. The optimal number of clusters to retain corresponds to the maximum average silhouette value, $\bar{S}_i$, across all individuals in the sample.

We can plot the values of $\bar{S}_i$ for a set of cluster solutions (e.g., 1 through 10) and then plot them using the R commands below. Note that the R function call is very similar to that for plotting WSS, where we simply indicate silhouette as the method.

```
fviz_nbclust(civics.cluster.data, kmeans, method="silhouette")
```

The function not only provides us with the graph but also includes a line at the maximum silhouette value, which is 2 in this case.

A third approach for determining the number of clusters to retain is the gap statistic (Tibshirani, Walther, & Hastie, 2001), which is calculated as:

$$GAP_k = \frac{1}{B}\sum_{b=1}^{B} ln\left(trW_k^{(b)} - ln(trW_k)\right) \tag{6.4}$$

where
  B = Number of reference datasets generated under the null distribution
  $W_k^{(b)}$ = Matrix of within-cluster sums of squares for reference dataset b in the k cluster solution
  W_k = Matrix of within-cluster sums of squares for observed data in the k cluster solution

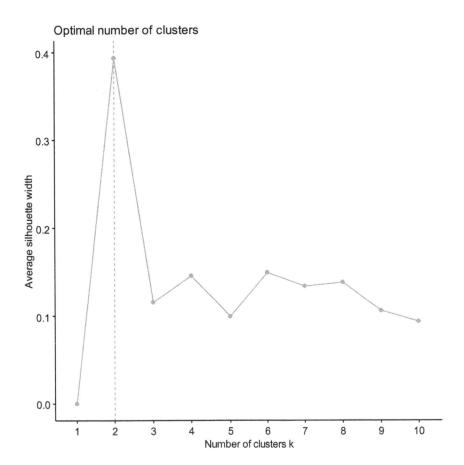

FIGURE 6.2
Silhouette value by number of clusters for K-means solution

We calculate GAP_k and the standard deviation of $trW_k^{(b)}$, s_k
    We then calculate

$$\hat{s}_k = s_k \sqrt{1 + 1/B} \tag{6.5}$$

The estimated number of clusters is the smallest k such that

$$GAP_k \geq \left(GAP_K^* - \tilde{s}_k^* \right)$$

where
    $k^* = \text{argmax } GAP_{k^*}.$

The reference distribution against which the observed within sums of squares are compared can be constructed in one of two ways:

1. Sample from the uniform distribution within each variable with the range of values falling between the min and max of observed values for that variable.
2. The variables are sampled from a uniform distribution over a box aligned with the principal components of the centered observed variables. The new set of variables is then back-transformed to obtain a reference dataset.

The major advantage of approach 2 is that it takes into account the shape of the original data distribution.

We can obtain and graph the gap statistic using the fviz _ nbclust function in R. The call is similar to those for the WSS and silhouette statistics, with the addition of the number of samples to use when creating the reference distribution; 1000 in this case.

```
fviz_nbclust(civics.cluster.data, kmeans, method="gap_
stat", nboot=1000)
```

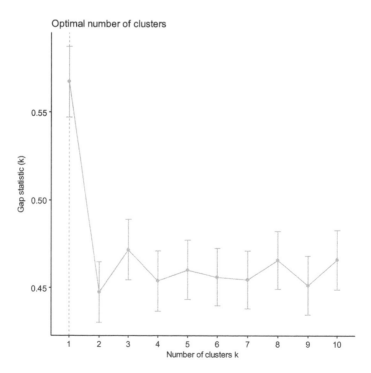

FIGURE 6.3
Gap statistic value by number of clusters for K-means solution

The optimal number of clusters based on the gap statistic is 1.

A fourth approach for determining the number of clusters to retain is the Clest statistic (Dudoit & Fridlyand, 2002). The steps involved for the Clest technique are as follows. Define the maximum number of clusters to be considered as M.

For each number of clusters $2 \leq k \leq M$ perform the following steps:

1. Repeat the following B times:
   a. Randomly split sample into 2 nonoverlapping sets L^b and T^b.
   b. Apply clustering procedure to L^b to obtain partition PL^b.
   c. Build a classifier CL^b using L^b and its cluster labels.
   d. Apply the CL^b to T^b to obtain CT^b.
   e. Apply the clustering procedure to T^b to obtain PT^b.
   f. Compute an external index s_{kb} comparing CT^b to PT^b.

2. Let $t_k = median(s_{kb})$ denote the observed similarity index (e.g., FM) for the k cluster solution.

3. Generate B_0 reference datasets under the null hypothesis. For each reference dataset, repeat steps 1 and 2 to obtain B_0 similarity index values t_{kB_0}.

4. Let t_k^0 be the average of the B_0 t_{kB_0} values.
   a. Let p_k denote the proportion of the t_{kB_0} values that are at least as large as the observed t_k value. This is the p-value for t_k.
   b. Let $d_k = t_k - t_{kB_0}$; difference between observed and null expected similarity statistic values. Here the null is $k = 1$.

In order to determine the number of clusters to retain, compare the following:

p_k to a threshold (p_{max}; e.g. 0.05); when it's less we have a significant effect.

d_k to a minimal distance threshold (d_{min}); when it's greater we have an important effect.

If neither is past its threshold, then $k=1$.

Otherwise, let $k=$the maximum of d_k where $p_k < 0.05$

The Clest statistic is available in the RSKC R library and can be calculated using the following command. We use the Clest function and save the results in an output object.

```
civics.cluster.clest0<-Clest(as.matrix(civics.cluster.
data), 10, alpha=0, L1=NULL, B=500,B0=100, beta = 0.05,
pca=TRUE, silent=TRUE)
civics.cluster.clest0$K
[1] 1
```

We need to provide the dataset name and coerce it to be a matrix object. We then specify the maximum number of clusters (10), the proportion of observations to be trimmed (0), the penalty tuning parameter (no tuning in this example), the number of bootstrap samples (1000), the number of reference datasets (100), the value of the threshold against which the bootstrapped *p*-values are compared, how to generate the reference distribution (PCA versus simple), and whether we want results for each bootstrap sample to be printed or not (silent=TRUE means that they will not be printed).

The results indicate that we should retain only a single cluster.

Finally, we can use the NbClust function from the NbClust library to obtain results for 30 indices often used to determine the number of factors to retain. This function uses a vote-counting approach where, for each of the indices, the number of clusters to be retained is determined and then the vote totals are presented. We would retain the number of clusters that have the highest number of votes. The function call and results using the NB approach appears below. When using the function, we need to supply the dataset name, the clustering distance metric, the minimum and maximum number of clusters to check, and the clustering method to use.

```
civics.nb <- NbClust(civics.cluster.data, distance =
"euclidean", min.nc = 2,max.nc = 10, method = "kmeans")

*** : The Hubert index is a graphical method of determining
the number of clusters.
                In the plot of Hubert index, we seek a
significant knee that corresponds to a
                significant increase of the value of the
measure i.e the significant peak in Hubert
                index second differences plot.


*** : The D index is a graphical method of determining the
number of clusters.
                In the plot of D index, we seek a
significant knee (the significant peak in Dindex
                second differences plot) that corresponds
to a significant increase of the value of
                the measure.
*******************************************************************
* Among all indices:
* 6 proposed 2 as the best number of clusters
* 5 proposed 3 as the best number of clusters
* 4 proposed 4 as the best number of clusters
* 1 proposed 5 as the best number of clusters
* 1 proposed 6 as the best number of clusters
* 3 proposed 9 as the best number of clusters
* 3 proposed 10 as the best number of clusters
```

***** Conclusion *****

* According to the majority rule, the best number of
clusters is 2

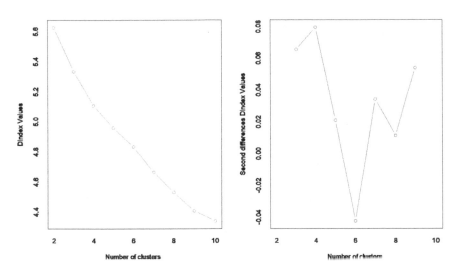

Taken together, the results provide us with a mixed set of results regarding the number of clusters to retain. The predominant result would suggest that there are likely 1 or 2 clusters in the data, although the WSS result suggested 5. For this reason, we will fit 2, 3, and 4 cluster solutions. We'll start with the 4 cluster solution and then print out the full set of results. When using the kmeans function, we need to specify the dataset name, the number of clusters (i.e., clusters), and the number of random starts.

```
kmeans.cluster4 <- kmeans(civics.cluster.data, centers=4,
nstart = 50)
kmeans.cluster4
K-means clustering with 4 clusters of sizes 4, 26, 49, 21
```

```
Cluster means:
  citizenship     social economy security    trust patriotism     women immigration   schoolA politicalA   climate
1    4.725000   6.185000 8.604701 9.942500  4.79750  8.097889  7.435057    6.360000  8.830794   9.322500  8.672531
2    9.636538 10.233652 8.799185 9.962385 10.05131 12.534291  9.198086    9.979825 10.127167   9.032505 10.079255
3   10.497295  8.937136 8.552303 9.141048 10.63455  8.711855  9.130450    9.691798  9.504869  11.141165 10.252270
4   12.697143 12.512381 9.791429 9.953810 11.42165 10.571026 10.273000   11.701471 10.616644  11.386597 11.423833
```

```
Clustering vector:
  [1]  2 1 3 3 3 2 1 3 3 3 3 2 2 3 3 2 1 3 4 3 2 3 3 3 3 3 3
 3 4 3 2 3 3 3 2 3 4 3 3 3 2 2 3 2 4 3 2 2 3 3 3 3 3 2 4
 [56]  3 4 4 3 4 4 2 3 2 2 2 4 4 2 4 1 4 3 4 2 3 4 3 2 3 2 2
 3 3 2 4 4 2 4 4 3 2 4 3 3 4 3 3 3 3
```

```
Within cluster sum of squares by cluster:
[1]   221.0440   803.2038 1184.4886   858.4737
 (between_SS / total_SS =   29.4 %)
```

The difference between clusters accounted for 29.4% of the total variance in the data. Clusters 1, 2, 3, and 4 contained 4, 26, 49, 21 observations, respectively. Based on the means, it appears that cluster 1 was characterized by low scores on the citizenship, social, trust, women, immigration, and climate measures. In contrast, cluster 4 had the highest means for citizenship, social, economy, trust, women, immigration, and climate. Clusters 2 and 3 had similar means for economy, security, trust, women, immigration, and climate, whereas cluster 2 had higher means for social, patriotism, and schoolA. Cluster 3 had higher means than cluster 2 for citizenship and politicalA.

Next, we can plot the clusters using the following R function, which is part of the factoextra library. We need to specify the cluster solution and the dataset.

```
fviz_cluster(kmeans.cluster4, data=civics.cluster.data)
```

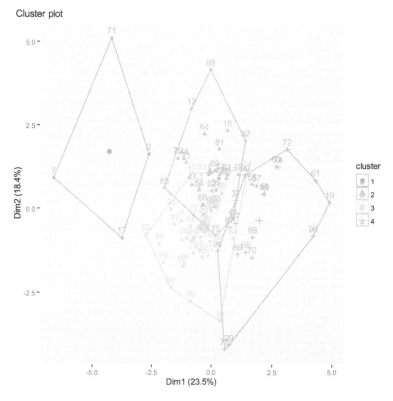

FIGURE 6.4
Cluster membership plot for *k*=4

The 2 dimensions of the graph represent the results of a principal components analysis of the variables used in the cluster analysis. The individual members of the sample are plotted in the graph, and the cluster borders are represented by lines connecting the individuals who are furthest from the cluster centers. From this graph, we can see that cluster 1 is completely separated from the other clusters, and that cluster 4 has relatively little overlap with clusters 2 and 3, which overlap with one another quite a bit.

Next, we'll fit the 3 cluster solution.

```
kmeans.cluster3 <- kmeans(civics.cluster.data, centers=3,
nstart = 50)
kmeans.cluster3
kmeans.cluster3
K-means clustering with 3 clusters of sizes 57, 4, 39
```

```
Cluster means:
  citizenship   social  economy  security    trust patriotism   women immigration   schoolA  politicalA   climate
1   10.81153  9.435082 8.467243  9.028094 10.70811   8.701474 9.042847    9.823922  9.550964  11.182259 10.268623
2    4.72500  6.185082 8.604701  9.942500  4.79750   8.097889 7.435057    6.360000  8.830794   9.322500  8.672531
3   10.64872 10.998845 9.508431 10.291333 10.56204  12.276410 9.918986   10.772843 10.451014   9.807486 10.743869
```

```
Clustering vector:
  [1] 3 2 1 1 1 3 2 1 1 1 1 3 3 1 1 3 2 1 3 1 3 1 1 1 1 1 1
 1 1 1 3 1 1 1 3 1 3 1 1 1 1 3 3 1 3 3 1 3 3 1 1 1 1 1 3 3
 [56] 1 3 3 1 3 3 3 1 3 3 3 1 1 3 3 2 3 1 3 3 1 1 1 3 1 3 3
 1 1 1 1 1 3 3 3 1 3 1 1 1 3 1 1 1 1
```

```
Within cluster sum of squares by cluster:
[1] 1558.283   221.044 1601.415
 (between_SS / total_SS =   22.2 %)
```

The 3 clusters accounted for 22.2% of the total variance in the data. Based on the pattern of means, cluster 2 in this solution corresponds to cluster 1 from the 4 cluster solution. In addition, cluster 3 had the highest values for social, economy, security, patriotism, immigration, and schoolA. Cluster 1 had the highest mean for politicalA, and comparable means to cluster 3 for citizenship, trust, women, and climate. Cluster 2 had the lowest mean values for citizenship, social, trust, women, immigration, and climate. We can plot the 3 cluster solution as we did earlier.

```
fviz_cluster(kmeans.cluster3, data=civics.cluster.data)
```

Cluster 2 is clearly distinct from the other two, whereas clusters 1 and 3 have some overlap, much as did clusters 2 and 3 in the 4 cluster solution.

Finally, we will fit the 2 cluster solution.

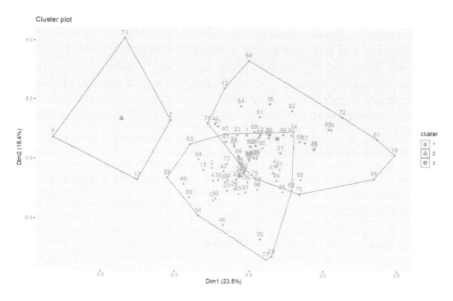

FIGURE 6.5
Cluster membership plot for *k*=3

```
kmeans.cluster2 <- kmeans(civics.cluster.data, centers=2,
nstart = 50)
kmeans.cluster2
K-means clustering with 2 clusters of sizes 58, 42
```

Cluster means:

	citizenship	social	economy	security	trust	patriotism	women	immigration	schoolA	politicalA	climate
1	10.35530	9.02965	8.499856	9.117609	10.25117	8.630069	8.919867	9.499795	9.459913	11.023225	10.16606
2	10.71071	11.13750	9.402114	10.164571	10.64058	12.062180	9.873112	10.822770	10.443874	9.948182	10.69955

```
Clustering vector:
  [1] 2 1 1 1 1 2 1 1 1 1 1 2 2 1 1 2 1 1 2 1 2 1 2 1 1 1 1 1 1 1
1 1 1 2 1 1 1 2 1 2 1 1 1 2 2 1 2 2 1 2 2 1 1 1 1 1 2 2
 [56] 1 2 2 1 2 2 2 1 2 2 2 1 2 2 2 1 2 1 2 2 1 1 1 2 1 2 2
1 1 1 2 1 2 2 2 1 2 2 1 1 2 1 1 1 1
```

```
Within cluster sum of squares by cluster:
[1] 2071.696 1700.045
 (between_SS / total_SS =  13.2 %)
```

When only 2 clusters were retained, 13.2% of the total variance in the variables was accounted for by the clusters. With respect to the clustering variables, cluster 2 had the highest means for social, economy, security, patriotism, women, immigration, and schoolA. Cluster 1 had the highest mean score politicalA, and the two groups had similar means for citizenship, trust, and climate. The cluster plot appears in Figure 6.6. Cluster 2 has two apparent outlying observations (7 and 71), and the clusters display no small amount of overlap.

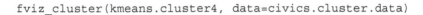

```
fviz_cluster(kmeans.cluster4, data=civics.cluster.data)
```

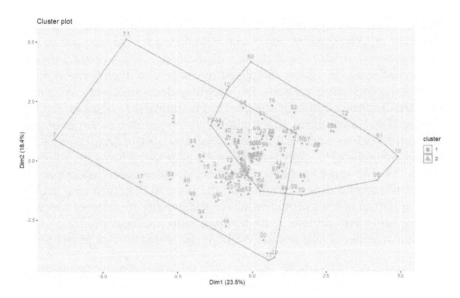

FIGURE 6.6
Cluster membership plot for *k*=4

Now that we have fit several cluster models, and have examined the statistics providing information regarding the number to retain, we will need to make a decision regarding which solution we believe to be optimal. It is important to note that this decision is, in the final analysis, subjective. There is no single statistic or analytic result that can provide a definitive answer to the question of how many clusters are present in the data. Of paramount importance is that the solution be substantively meaningful and defensible. Based on this criterion, it would appear that the 2 cluster solution is not optimal. The clusters overlap with one another, and there are outliers present in cluster 2. With respect to choosing between cluster solutions 3 and 4, we need to decide how theoretically meaningful cluster 4 is. Clearly, it is separate from the others, based on the cluster graph. Thus, the question for us to consider is whether the pattern of means for this cluster (highest on citizenship, social, economy, trust, women, immigration, and climate) corresponds to a conceptually meaningful subgroup within the population, or not. In this case, cluster 4 does appear to be theoretically defensible, and thus we will retain the 4 cluster solution.

Regularized K-means Cluster Analysis

K-means clustering can have difficulty finding the correct solution when the number of variables, p, is large and/or many of the variables used in the analysis are not relevant to the problem of differentiating the subgroups in the population (Witten & Tibshirani, 2010). Such a scenario could occur, for example, when the researcher is unsure of which variables will best differentiate the underlying subgroups, and so includes a large number of them in the cluster analysis. The inclusion of many variables could introduce noise with respect to cluster membership, thereby reducing the accuracy of traditional clustering methods. In order to address this issue, Witten and Tibshirani described a variant of K-means clustering that relies on the lasso estimator (Tibshirani, 1996). In situations where regularization may prove to be useful, such as with high-dimensional data and/or small samples, we may find it beneficial to apply the lasso estimator to the K-means clustering problem. As with other applications of this estimator described in previous chapters, the central aspect of applying the lasso to the clustering problem is determination of the optimal value for the penalty parameter. Once this value has been determined, it can be applied to the basic problem of estimating the number of clusters that best approximates what is present in the data.

In the context of cluster analysis, the lasso estimator (SPARCL) is designed to maximize the following function:

$$\sum_{j=1}^{p} w_j f_j\left(x_j \Theta\right) \tag{6.6}$$

where

x_j = Variable j
w_j = Weight indicating the contribution of variable j to the cluster solution
Θ = Cluster partition consisting of number of clusters and individual cluster membership
f_j = Function indicating degree of cluster separation; e.g., between cluster sum of squares

Larger values of w_j indicate that the variable contributes relatively more to the cluster solution, whereas when $w_j = 0$ the variable does not contribute to the solution at all. The value of w_j is determined in part by the sparsity parameter s, which the researcher must select based on statistics reflecting relative model fit, as described in the K-means section above. The values of the weights are subject to the following:

$$\|w\|^2 \leq 1, \|w\|_1 \leq s, w_j \geq 0 \forall_j \tag{6.7}$$

where
  w = Vector of variable weights

We can apply the lasso penalty to the K-means clustering problem using functions in the `sparcl` R library. As was true with other applications of regularization methods that are described in this book, our first task is to identify the optimal value of the lasso tuning parameter. This can be done using the following command. Note that we will need to identify the optimal tuning parameter value separately for each number of clusters. The function for this purpose is `KMeansSparseCluster.permute`, which makes use of a permutation-based approach to assess multiple tuning parameter values, and works as follows:

1. Fit a K-means solution to data using tuning parameter value s to obtain an objective function value measuring cluster separation $O(s)$; e.g., between cluster sum of squares.
2. Randomly permute values for each variable and fit cluster solution to data using tuning parameter s to obtain $O^*(s)$.
3. Repeat step 2 B times.
4. Calculate the Gap statistic as $Gap(s) = ln(O(s)) - mean(ln(O^*(s)))$
5. Repeat steps 2–4 for several values of s.
6. The optimal value of s corresponds to the maximum value of $Gap(s)$.

In short, the optimal value of s is the one that maximizes the difference between the observed cluster separation value and that for the data under the null hypothesis of no coherent clusters, as represented by the randomly permuted data.
  The function call for conducting the permutation analysis takes the following form:

```
sparcl.perm2<- KMeansSparseCluster.permute(civics.cluster.
data, K=2, nperms=100, nvals=100)

sparcl.perm2$bestw
[1] 1.352538
```

The results of the analysis are saved in an output object. We need to specify the dataset, the number of clusters, the number of permutations, and the number of s values to check. The optimal tuning parameter is saved in the object `sparcl.perm2$bestw`. In this example, we request 100 permutations and 100 values of s. For the 2 cluster solution the optimal tuning parameter value is 1.352538.

```
plot(sparcl.perm2)
```

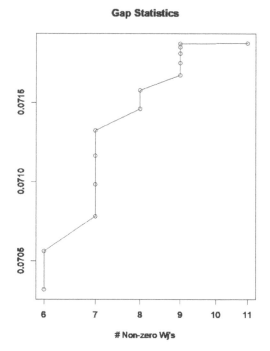

FIGURE 6.7
Gap statistic by number of non-zero weights for optimal value of s for the 2 cluster solution

Once the optimal value of *s* is determined for the 2 cluster solution, we can use it to fit K-means for 2 clusters using the KMeansSparseCluster function. We need to specify the dataset, the number of clusters, and the value of the tuning parameter, which is contained in sparcl.perm2$bestw from the permutation analysis output.

```
sparcl.results2 <- KMeansSparseCluster(civics.cluster.
data,K=2,wbounds=sparcl.perm2$bestw)
```

By typing in the name of the output object, we obtain the cluster assignments for each individual in the sample appearing after the Clustering: label.

```
Sparcl.results2
Wbound is  1.352538 :
Number of non-zero weights:   11
Sum of weights:  1.347756
Clustering:  1 2 2 2 2 1 2 2 2 2 2 1 1 2 2 2 2 2 1 2 1 2 2
2 2 2 2 2 2 1 2 2 2 2 1 2 1 2 1
1 2 1 1 2 1 1 2 2 2 2 2 1 1 2 2 2 2 1 2 1 2 1 1 1 2 2 1 1 2
1 2 1 1 2 2 2 1 2 1 1 2 1 2 2 2 1
2 1 2 1 2 2 2 1 2 2 2 2
```

We can plot the sparse clustering solution, which yields the variable weight on the *y*-axis, and the variable index (in order of how they appear in the dataset) on the *x*-axis.

```
plot(sparcl.results2)
```

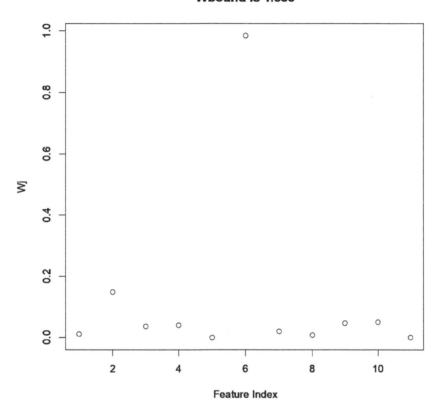

Wbound is 1.353

FIGURE 6.8
Variable weights by variable index number for optimal value of s with the 2 cluster solution

From this graph, we can see that variable 6 (patriotism) has the largest weight, followed by variable 2 (social). All of the other variables had weights at or near 0. We can see the actual weight values by using the following command sequence to save the weights for the 2 cluster solution in the object sparcl.weights2.

```
sparcl.classification2<-unlist(lapply(sparcl.results2,
  '[[', 2))
sparcl.weights2<-unlist(lapply(sparcl.results2, '[[', 1))
```

```
sparcl.weights2
citizenship        social      economy      security      trust    patriotism           women
1.154974e-02 1.485151e-01 3.640065e-02 4.041065e-02 8.829047e-06 9.847071e-01 2.036522e-02
 immigration        schoolA    politicalA        climate
8.125306e-03 4.694039e-02 5.019250e-02 5.407793e-04
```

Again, we see that the 2 cluster solution was driven primarily by the patriotism variable, followed distantly by social, with the other variables playing little to no role in the solution. We can obtain the variable means by cluster using the `aggregate` function, as follows. Cluster 1 had higher means on both social and patriotism.

```
aggregate(civics.cluster.data, by=list(sparcl.
classification2), mean)
Group.1 citizenship     social economy security      trust patriotism    women immigration
1        1    10.78800 10.931284 9.381966 10.087486 10.42256  12.688571 9.696585   10.293168
2        2    10.35196  9.367687 8.607871  9.271867 10.41050   8.662393 9.117577    9.927439
    schoolA politicalA   climate
1 10.44456   9.980864 10.45145
2  9.56551  10.889854 10.35710
```

The cluster frequencies can be obtained using the `table` command.

```
table(sparcl.classification2)
sparcl.classification2
 1  2
35 65
```

The command sequence for fitting solutions with other numbers of clusters are similar for the 2 cluster solution, with a change to the K= subcommand to reflect the number of clusters. The command sequence presented above for the 2 cluster solution appears as follows for 3 clusters:

```
#3 clusters#
sparcl.perm3<- KMeansSparseCluster.permute(civics.cluster.
data, K=3, nperms=100, nvals=100)

sparcl.perm3$bestw
[1] 1.482945

plot(sparcl.perm3)

sparcl.results3 <- KMeansSparseCluster(civics.cluster.
data,K=3,wbounds=sparcl.perm3$bestw)

sparcl.results3
Wbound is  1.482945 :
Number of non-zero weights:   4
Sum of weights:  1.482941
```

Gap Statistics

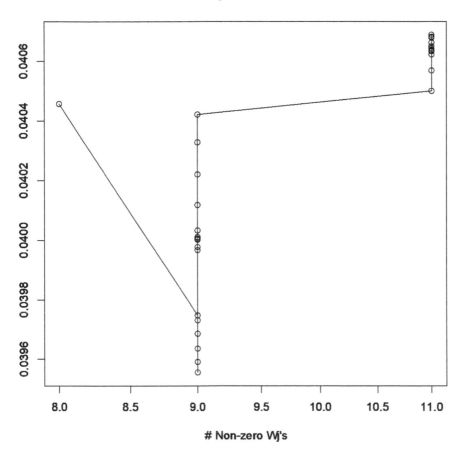

FIGURE 6.9
Gap statistics by number of non-zero weights for optimal value of s with the 3 cluster solution

```
Clustering:  1 2 3 3 3 1 2 3 3 3 3 3 1 1 3 3 3 2 3 1 3 1 3 3
3 3 3 3 3 3 3 1 3 3 3 3 3 1 3 1 3 1
1 3 1 1 3 1 1 3 3 3 3 3 1 1 3 3 3 3 1 3 1 3 1 1 1 3 3 1 1 2
1 3 1 1 3 3 3 1 3 1 1 3 1 3 3 3 1
3 1 3 1 3 3 3 1 3 3 3 3

plot(sparcl.results3)

sparcl.weights3
citizenship      social     economy     security       trust  patriotism       women
  0.2263217   0.1001641   0.0000000   0.0000000   0.2107694   0.9456859   0.0000000
immigration      schoolA   politicalA      climate
  0.0000000   0.0000000   0.0000000   0.0000000
```

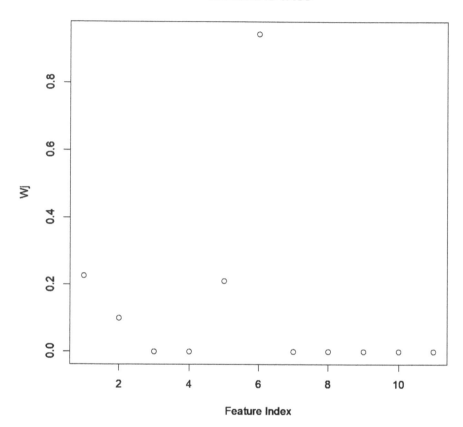

FIGURE 6.10
Variable weights by variable index number for optimal value of s with the 3 cluster solution

```
aggregate(civics.cluster.data, by=list(sparcl.
classification3), mean)
```

	Group.1	citizenship	social	economy	security	trust	patriotism	women	immigration
1	1	10.78800	10.931284	9.381966	10.087486	10.42256	12.688571	9.696585	10.29317
2	2	4.72500	6.185000	8.604701	9.942500	4.79750	8.097889	7.435057	6.36000
3	3	10.72094	9.576388	8.608079	9.227891	10.77857	8.699410	9.227907	10.16137

```
     schoolA politicalA    climate
1 10.444558   9.980864 10.451454
2  8.830794   9.322500  8.672531
3  9.613688  10.992631 10.467566
```

When 3 clusters are retained, the patriotism variable remains the most important, with citizenship, trust, and social also having non-zero weights. The other variables' weights were regularized to 0, meaning that they were not involved in the cluster solution at all. The aggregated means show that cluster 2 had the lowest means on citizenship, social, and trust, whereas cluster 1 had the highest means on citizenship (along with cluster 3), social, trust (with cluster 3), and patriotism. Cluster 3 had the largest means, along with cluster 1, on citizenship and trust, and the second highest means on the other variables which weights that were not regularized to 0. The cluster frequencies appear as follows:

```
table(sparcl.classification3)
sparcl.classification3
 1  2  3
35 22 43
```

Results for the 4 cluster solution appear as follows:

```
#4 clusters#
sparcl.perm4<- KMeansSparseCluster.permute(civics.cluster.
data, K=4, nperms=100, nvals=100)

sparcl.perm4$bestw
[1] 1.211097

plot(sparcl.perm4)

sparcl.results4 <- KMeansSparseCluster(civics.cluster.
data,K=4,wbounds=sparcl.perm4$bestw)

sparcl.results4
Wbound is  1.211097 :
Number of non-zero weights:  2
Sum of weights:  1.211103
Clustering:  1 2 2 4 4 4 2 4 4 3 2 1 1 3 3 2 3 3 4 3 1 3 4
3 3 3 4 4 3 3 1 3 2 3 3 3 4 3 4 2 1
1 3 1 1 3 1 1 3 4 3 4 3 4 1 3 3 4 3 1 4 4 3 1 1 1 3 3 1 1 2
1 3 4 1 4 3 4 1 3 4 1 4 4 2 3 3 1
4 4 4 1 4 3 3 1 3 3 4 3

plot(sparcl.results4)

sparcl.classification4<-unlist(lapply(sparcl.results4,
'[[', 2))
sparcl.weights4<-unlist(lapply(sparcl.results4, '[[', 1))
sparcl.weights4
citizenship      social     economy    security      trust patriotism      women
 0.2404392   0.0000000   0.0000000   0.0000000   0.0000000   0.9706642   0.0000000
immigration      schoolA   politicalA     climate
 0.0000000   0.0000000    0.0000000    0.0000000

aggregate(civics.cluster.data, by=list(sparcl.
classification4), mean)
  Group.1 citizenship     social economy  security     trust patriotism    women immigration
1       1   10.679200  10.796667 9.170352 10.236080 10.231178  13.220000 9.662418   10.102836
2       2    6.056667   7.396667 8.959867  9.980000  9.154969   8.870346 8.333359    8.744444
```

Gap Statistics

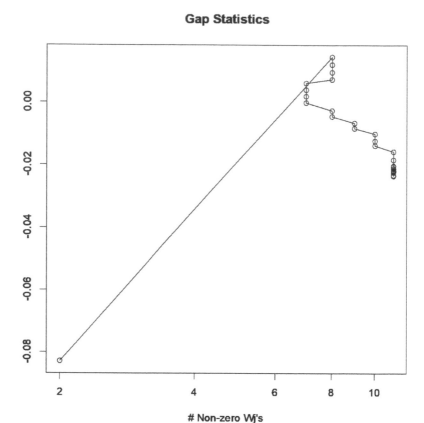

FIGURE 6.11
Gap statistics by number of non-zero weights for optimal value of s with the 4 cluster solution

```
3       3   11.080000  9.733684 8.460000  9.043684 10.483421   7.988158 9.011854   9.991842
4       4   10.997409 10.183560 9.160815  9.512548 10.890277  10.474016 9.750423  10.520841
       schoolA politicalA  climate
1 10.434254    9.772341 10.30082
2 10.246261    8.810000 10.51779
3  9.302799   11.346187 10.19870
4 10.026378   10.800610 10.68862
```

When 4 clusters are retained, only the patriotism and citizenship variables had non-zero weights. Cluster 2 had the lowest mean on citizenship, whereas clusters 3 and 4 had the highest means, followed closely by cluster 1. For patriotism, cluster 1 had the highest mean, followed by cluster 4, cluster 2, and finally cluster 3. The number of individuals in each cluster appear below using the table command.

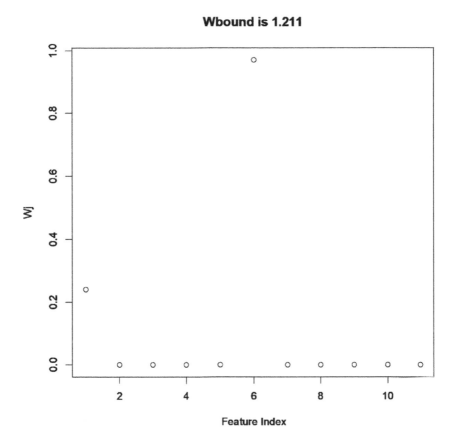

FIGURE 6.12
Variable weights by variable index number for optimal value of s with the 4 cluster solution

```
table(sparcl.classification4)
sparcl.classification4
 1  2  3  4
25  9 38 28
```

Finally, results for the 5 cluster solution appear as follows:

```
#5 clusters#
sparcl.perm5<- KMeansSparseCluster.permute(civics.cluster.
data, K=5, nperms=100, nvals=100)

sparcl.perm5$bestw
[1] 1.883915

plot(sparcl.perm5)
```

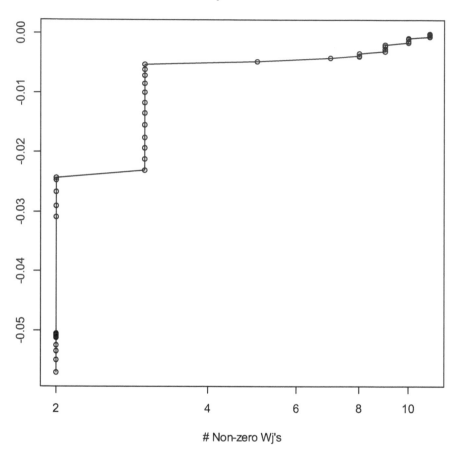

FIGURE 6.13
Gap statistics by number of non-zero weights for optimal value of s with the 5 cluster solution

```
sparcl.results5 <- KMeansSparseCluster(civics.cluster.
data,K=5,wbounds=sparcl.perm5$bestw)

sparcl.results5
Wbound is  1.883915:
Number of non-zero weights:   6
Sum of weights:  1.883893
Clustering:   1 2 2 3 3 1 4 3 3 3 2 1 1 2 2 2 2 3 5 3 1 3 2
2 2 3 3 3 5 3 1 3 2 2 3 3 1 2 3 2 1 1 3 1 1 3 1 1 2 3 3
3 2 1 1 3 3 3 2 5 3 1 3 1 1 1 3 3 1 5 4 1 3 1 1 3 5 3 1 3 1
1 3 3 1 3 3 1 3 3 3 1 3 2 3 5 2 3 2 2

plot(sparcl.results5)
```

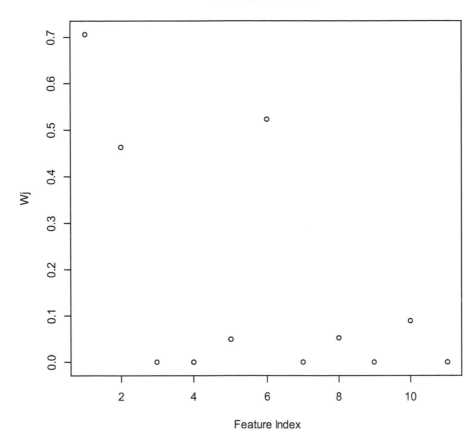

FIGURE 6.14
Variable weights by variable index number for optimal value of s with the 5 cluster solution

```
sparcl.classification5<-unlist(lapply(sparcl.results5,
'[[', 2))
sparcl.weights5<-unlist(lapply(sparcl.results5, '[[', 1))

sparcl.weights5
citizenship     social    economy   security     trust patriotism    women immigration    schoolA
0.70658358 0.46301382 0.00000000 0.00000000 0.04899049 0.52254807 0.00000000  0.05251720 0.00000000
politicalA     climate
0.09023984  0.00000000

aggregate(civics.cluster.data, by=list(sparcl.
classification5), mean)
    Group.1 citizenship     social economy security     trust patriotism    women immigration   schoolA politicalA
1       1    9.857241 10.619136 9.026166 10.009034 10.192557  12.755226 9.495870   10.210188 10.315501    9.443457
```

```
2    2    9.269524  7.472857 8.579943  9.561429  9.715714    8.015714 9.388571   10.433333  9.408246 10.892938
3    3   11.112082 10.401183 8.838401  9.148127 10.894232    9.198630 9.268388    9.935085  9.799728 11.107114
4    4    1.780000  4.020000 8.720000 10.025000  6.135000    9.080778 6.175115    4.235000  9.141588  6.980000
5    5   16.611667 13.620000 9.548333 10.068333 12.005000   10.736667 9.643372   10.767574 10.120529 12.350000
     climate
1 10.146582
2  9.719530
3 10.883366
4  9.095062
5 10.893333
```

```
table(sparcl.classification5)
sparcl.classification5
 1  2  3  4  5
29 21 42  2  6
```

When we retain 5 clusters, it is important to note that cluster 4 contains only 2 individuals and cluster 5 has only 6 subjects, calling into question the utility of this solution in practice. Nonetheless, taking it at face value, we see that several variables have non-zero weights, with the largest belonging to citizenship, followed by patriotism and social. The weights for trust, immigration, and politicalA are quite a bit lower than those of citizenship, patriotism, and social, suggesting that relatively speaking, they do not contribute much to the final cluster solution. Cluster 5 exhibited the highest scores for citizenship and social, and the second-highest mean for patriotism. Cluster 4 had the lowest means for citizenship and social, and cluster 2 had the lowest patriotism mean. Cluster 3 had the second-highest means for citizenship, social, and patriotism, with cluster 1 having the highest patriotism mean.

Determining the Number of Clusters to Retain

Given this set of solutions, we must now determine which is optimal. Results of the KMeansSparseCluster.permute command yield the gap statistic for each value of the tuning parameter for each cluster solution. Therefore, we can use these values to identify the number of clusters that should be retained. Recall that the solution associated with the largest gap statistic is considered to be optimal and that the optimal sparse cluster solution is selected by maximizing gap as well. Thus, we can compare the max gaps from the permutation results in order to determine which number of clusters to retain. The gap statistics for each cluster solution for each shrinkage tuning parameter value are saved in an object labeled gaps. We can obtain the maximum gap statistic value for each cluster solution using the following commands:

```
max(sparcl.perm2$gaps)
[1] 0.1156652
```

```
max(sparcl.perm3$gaps)
[1] 0.03857706
max(sparcl.perm4$gaps)
[1] 0.01195296
max(sparcl.perm5$gaps)
[1] -2.416135e-05
```

On the basis of these results, we would conclude that based on the gap statistic, 2 clusters were optimal.

In addition to the Gap statistic, we can also apply the Clest approach to the problem of determining the optimal number of clusters to retain using the tuning parameters that we obtained using the sparcl library. As we described earlier in the chapter, Clest is available through the RSKC R library. Following is the command sequence for doing this, followed by the associated output. The function call is quite similar to that which we used with the standard K-means examples discussed earlier in the chapter. Note that we are using the optimal penalty parameter that was associated with the 2 cluster solution. However, we can apply various values here (such as the optimal values associated with other cluster solutions) in order to fully understand the likely number of clusters present in the data. Also, note that we set alpha=0 for the lasso estimator, and alpha=1 for the ridge.

```
civics.cluster.clest2<-Clest(as.matrix(civics.cluster.
data), 5, alpha=0, L1=1.3157, B=1000,B0=10, beta = 0.05,
nstart=100, pca=FALSE, silent=TRUE)

civics.cluster.clest2$K
[1] 1

civics.cluster.clest2$result.table
    test.stat     obsCER     refCER P-value
2 0.3662879 0.4772727 0.11098485       1
3 0.4113636 0.4734848 0.06212121       1
4 0.3054924 0.4318182 0.12632576       1
5 0.2320076 0.3844697 0.15246212       1
```

These results show that based on the Clest penalty, there is 1 cluster present in the data. Examples of using Clest with other regularization values appear in the R script associated with this chapter, and which appears at the book website www.routledge.com/9780367408787.

Hierarchical Cluster Analysis

The K-means clustering approach begins with an *a priori* determination of the number of clusters to retain. Obviously, multiple such solutions can be fit to the data, but for any single analysis, the researcher must specify the number of clusters that will be assumed to exist. An alternative approach is known as hierarchical cluster analysis (HCA) and does not require an *a priori* specification for the number of clusters. In particular, we will focus on agglomerative clustering, in which every observation is assumed to be a unique cluster unto itself and then are grouped together based on a measure of similarity, such as Euclidean distance. More specifically, HCA works using the following algorithm:

1. Place each observation in a unique cluster of its own.
2. Combine 2 observations with the smallest distance between them.
3. Calculate between each observation and the new cluster.
4. Combine 2 observations/clusters with the smallest distance between them.
5. Repeat steps 2–4 until all observations are in a single cluster.

Perhaps the most commonly used measure of similarity is the Euclidean distance, which is calculated as:

$$d_{p,q} = \sqrt{\sum_{j=1}^{J}\left(q_j - p_j\right)^2} \tag{6.8}$$

where
   q_j = Value of variable j for individual q
   p_j = Value of variable j for individual p

There are several techniques for deciding how observations/clusters should be combined with one another in the algorithm described earlier. Some of the more popular of these include the following:

1. Single linkage (nearest neighbor)—Distance between clusters is defined as the distance between the two closest observations in each cluster.
2. Complete linkage—Distance between clusters is defined as the distance between the two furthest observations in each cluster.
3. Average linkage—Distance between clusters is defined as the average distance between all observations in the two clusters.
4. Centroid linkage—Distance between clusters is defined as the distance between the multivariate means (centroids) of the two clusters.
5. Ward's method—Distance between clusters is defined as the sum of squares within clusters summed across all variables. Combine clusters that minimize the within-cluster sum of squares for the resulting cluster solution.

Researchers may use multiple such approaches in practice, but perhaps the most popular is Ward's method (Hahs-Vaughn, 2017).

In R, we can use the `hclust` function, available in base R, to conduct HCA. We will need to make use of the `factoextra` R library in order to produce some of the graphs and statistics that will help us interpret the clustering results. First, we need to calculate the distances among the observations using the `dist` function.

```
#Calculate distances among observations#
civics.dist<-dist(civics.cluster.data, method="euclidean")
```

Next, we will obtain the HCA solution using the `hclust` function with the distance matrix created earlier and Ward's method. The results are saved in the object `civics.hca`.

```
civics.hca<-hclust(civics.dist, method="ward.D")
```

In order to obtain a dendogram, a graph displaying the results of the clustering algorithm, we use the following function, which is part of the `factoextra` library.

```
fviz_dend(civics.hca)
```

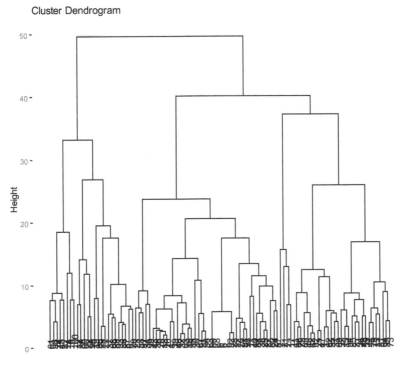

FIGURE 6.15
Dendogram for hierarchical cluster analysis

The individual observations appear at the bottom of the plot, and the lines representing the joining of these points and the resulting clusters. The length of the lines reflects the within-cluster variability for the joined clusters, so that long lines indicate greater variability in the newly combined cluster.

Ideally, we would be able to examine the structure of the dendogram to determine the number of clusters to retain. However, in many cases, this structure isn't particularly clear, so that alternative approaches for determining the number of clusters to retain must be considered. We can visualize varying numbers of cluster solutions using the `fviz _ dend` R function. For example, examining the dendogram here, we may consider the possibility that there are perhaps 3 or 4 clusters. Let's visualize the 3 cluster solution first. In the function call, we specify the number of clusters to retain, based on within-cluster variability.

```
fviz_dend(civics.hca, k=3, color_labels_by_k=TRUE, rect=TRUE)
```

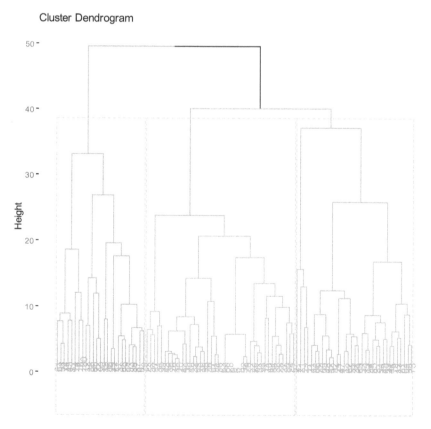

FIGURE 6.16
Dendogram for hierarchical cluster analysis with 3 factors retained

Though our interpretation is certainly subjective in nature, the 3 cluster solution appears to be clean, with the groups being distinct from one another. The dendogram for the 4 cluster solution appears as follows:

```
fviz_dend(civics.hca, k=4, color_labels_by_k=TRUE, rect=TRUE)
```

The 4 cluster solution is quite similar to that for 3 clusters, with the addition of a small group of individuals in cluster 3 (light blue).

A number of statistical indices have been proposed in an attempt to objectively identify the number of clusters to retain from HCA. Recall that we can obtain those using the NbClust function from the NbClust library. For HCA with Ward's method, we would use the following command sequence:

```
civics.nb.hca <- NbClust(civics.cluster.data, distance =
"euclidean", min.nc = 2, max.nc = 10, method = "ward.D")
```

```
*** : The Hubert index is a graphical method of determining
the number of clusters.
                In the plot of Hubert index, we seek a
significant knee that corresponds to a
                significant increase of the value of the
```

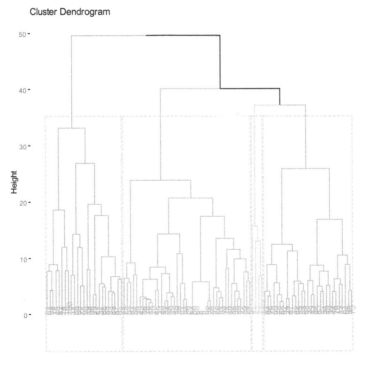

Cluster Dendrogram

FIGURE 6.17
Dendogram for hierarchical cluster analysis with 4 factors retained

```
measure i.e the significant peak in Hubert
                index second differences plot.

*** : The D index is a graphical method of determining the
number of clusters.
                In the plot of D index, we seek a
significant knee (the significant peak in Dindex
                second differences plot) that corresponds
to a significant increase of the value of
                the measure.

*************************************************************
* Among all indices:
* 9 proposed 2 as the best number of clusters
* 3 proposed 3 as the best number of clusters
* 1 proposed 4 as the best number of clusters
* 6 proposed 6 as the best number of clusters
* 2 proposed 7 as the best number of clusters
* 1 proposed 8 as the best number of clusters
* 2 proposed 10 as the best number of clusters

                ***** Conclusion *****

* According to the majority rule, the best number of
clusters is  2
```

The solution suggests that only 2 clusters should be retained, based on the set of indices. Indeed, the 2 cluster solution was by far the most common when judged by the number of indices suggesting it. We will conclude this section by examining the dendogram with the 2 cluster solution highlighted.

```
fviz_dend(civics.hca, k=2, color_labels_by_k=TRUE, rect=TRUE)
```

Subjectively, it would appear that the second cluster is actually made up of 2 separate subgroups, based on the shape of the dendogram. However, given that the statistical indices suggest the presence of 2 clusters, we must take this possibility seriously. We would next want to consider the theoretical/conceptual reasonableness of the cluster solutions with respect to the variable means and perhaps external criteria. It is important to keep in mind that whatever solution we finally decide upon must be theoretically sound.

Regularized Hierarchical Cluster Analysis

We can apply the regularized estimator to fitting HCA using functions from the sparcl library. The penalized estimator described earlier can be applied

Cluster Dendrogram

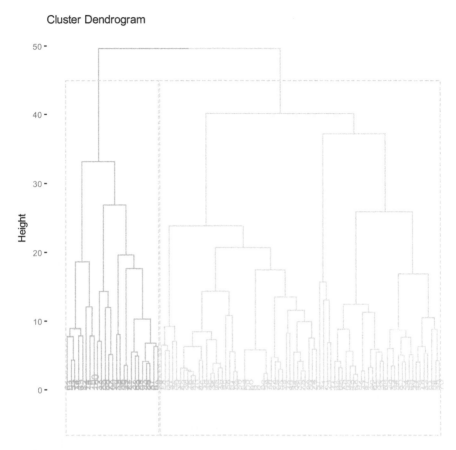

FIGURE 6.18
Dendogram for hierarchical cluster analysis with 2 factors retained

to the problem of HCA, much as it was for K-means. First, we need to identify the optimal tuning parameter value, which can be done using the following function call as applied to the civics data.

```
hca.perm<- HierarchicalSparseCluster.permute(as.
matrix(civics.cluster.data), wbounds=c(1.1:10),
nperms=100)
```

When calling the function, we need to convert our dataset to a matrix, provide the bounds within which the algorithm should search for the optimal tuning parameter, and the number of permutations to be used. We can elect not to provide the search bounds for the tuning parameter, in which case the function will select its own bounds for searching. We can print the candidate tuning parameters, along with the number of variables with non-zero weights, and the gap statistic (along with its standard deviation). In this case, the optimal tuning parameter was determined to be 4.1, which we can then use to determine the optimal regularized HCA solution. Also note that for this tuning parameter value, all 11 variables had non-zero weights.

```
print(hca.perm)
Tuning parameter selection results for Sparse Hierarchical
Clustering:
   Wbound # Non-Zero W's Gap Statistic Standard Deviation
1    1.1              2       0.0223             0.0005
2    2.1              9       0.1086             0.0023
3    3.1             11       0.1240             0.0024
4    4.1             11       0.1240             0.0024
5    5.1             11       0.1240             0.0024
6    6.1             11       0.1240             0.0024
7    7.1             11       0.1240             0.0024
8    8.1             11       0.1240             0.0024
9    9.1             11       0.1240             0.0024
Tuning parameter that leads to largest Gap statistic:  4.1
```

The following command will use the optimal tuning parameter determined by the permutation approach to conduct a regularized HCA, saving the results in the object sparse.hca.

```
sparse.hca <- HierarchicalSparseCluster(as.matrix(civics.
cluster.data), wbound=hca.perm$bestw, method="complete")
```

We need to specify the dataset to be clustered, the regularization parameter, and the agglomeration method. We can then print the dendogram for our regularized solution.

```
plot(sparse.hca$hc)
```

The dendogram suggests that the majority of the observations belong to a common cluster, with a small number appearing to belong to perhaps one or two additional clusters.

We can use the fviz _ dend function to explore the number of clusters that might be present, just as we did for standard HCA. The approach is similar to that applied with the standard HCA, as described earlier.

```
fviz_dend(sparse.hca$hc, k=2, color_labels_by_k=TRUE)
```

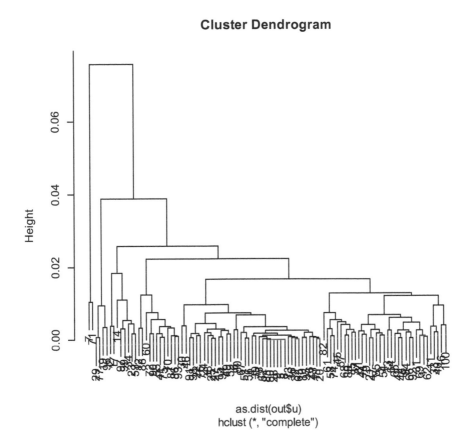

FIGURE 6.19
Dendogram for regularized hierarchical cluster analysis

This graph suggests that there is one dominant cluster, with a very small number of observations belonging to a second cluster. The dendogram highlighting 3 clusters appears as follows:

```
fviz_dend(sparse.hca$hc, k=3, color_labels_by_k=TRUE)
```

Once again, there appears to be a dominant single cluster, with two very small additional clusters. Finally, the dendogram with 4 clusters highlighted by color appear suggests that there is a single very large cluster, with small additional clusters.

```
fviz_dend(sparse.hca$hc, k=4, color_labels_by_k=TRUE)
```

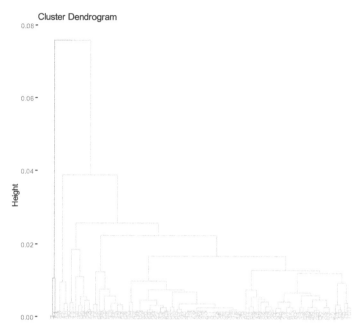

FIGURE 6.20
Dendogram for regularized hierarchical cluster analysis with 2 cluster solution

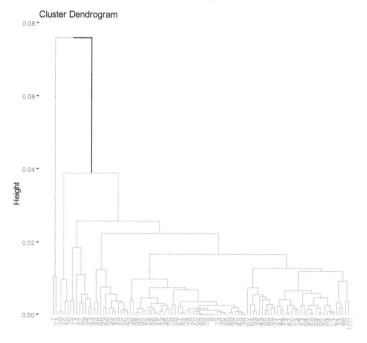

FIGURE 6.21
Dendogram for regularized hierarchical cluster analysis with 3 cluster solution

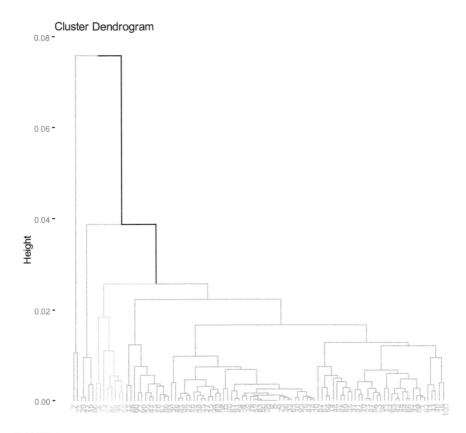

FIGURE 6.22
Dendogram for regularized hierarchical cluster analysis with 4 cluster solution

Finally, we will want to examine the weights applied to the variables in the clustering after regularization has been carried out using the optimal tuning parameter. The following code takes the names of the variables in the dataset and merges them with the HCA weights before printing them out:

```
cbind(civics.cluster.data.vars, sparse.hca$ws)
        [,1]            [,2]
 [1,]  "citizenship"  "0.621626022550657"
 [2,]  "social"       "0.408040087723967"
 [3,]  "economy"      "0.163444151291423"
 [4,]  "security"     "0.201985296533202"
 [5,]  "trust"        "0.272263991668164"
 [6,]  "patriotism"   "0.266202734855455"
 [7,]  "women"        "0.206437680231438"
 [8,]  "immigration"  "0.316531587824244"
 [9,]  "schoolA"      "0.142750951840286"
[10,]  "politicalA"   "0.1998076347568"
[11,]  "climate"      "0.177400845598867"
```

The variables with the largest weights were citizenship, social, and immigration. None of the variables had weights penalized to 0; i.e., all of the variables were involved in the clustering algorithm.

Summary

In this chapter, we have focused our attention on the application of regularization techniques, in particular the lasso, to the problem of CA. Clustering is a very widely used and popular statistical technique used for identifying subgroups within the larger population using multiple variables. Regularization can be applied to each of the two major types of CA, K-means and HCA. In addition, commonly used tools for determining the number of clusters such as the Gap and Clest statistics can also be used for identifying the optimal number of clusters to retain, and the optimal regularization tuning parameter, in the case of Gap. As with other statistical methods that we consider in this book, regularization is particularly useful for researchers who need to conduct CA in the context of high-dimensional data.

In the next chapter, we will turn our attention to regularization models for latent variables. In particular, our focus will be on both exploratory and confirmatory analysis and structural equation modeling. These methods are extensions of the univariate and multivariate models described in Chapters 3 and 5. We will begin these discussions with a review of the standard (non-regularized) models, followed by their regularized counterparts. And, as is the case with the other methods described in this book, we will demonstrate how these methods can be applied using the R software package.

7

Regularization Methods for Latent Variable Models

In the preceding chapters, we have examined a wide variety of statistical models designed for use with variables of different types, including those that are continuous and normally distributed, as well as those that are categorical in nature. We have also examined data analysis frameworks in which there is a single dependent variable (e.g., linear regression, logistic regression), multiple dependent variables (e.g., multivariate regression), and no dependent variables (e.g., canonical correlation). For each of these modeling frameworks, we have seen that regularization techniques are available and can be used to identify only those variables that are most salient in terms of the relationships involved. For the most part, these models involve only observed variables for which we can make direct measurements. Perhaps the exceptions to this assertion are canonical correlation and discriminant analysis, both of which center on the identification of a few linear combinations of the observed variables. However, even in those instances, the linear combinations are used to investigate relationships among sets of observed variables.

In Chapter 7, we will change course dramatically by examining latent variable models, and ways in which regularization techniques can be applied to them. Latent variables in this context refer to constructs that cannot be directly observed, but which we believe exist and that exert some influence on variables that can be directly measured. A classic example of this dynamic is items on a questionnaire or test. The items are written to measure one or more unobserved (latent) variables about which we would like to draw some conclusions. Examples of this latent structure abound in education and psychology, where researchers and practitioners are interested in latent variables such as cognitive proficiency, anxiety, depression, personality, and reading achievement, to name just a few. There exist a variety of statistical models for examining such latent variables, perhaps most importantly factor analysis. In this chapter, we will describe how regularization techniques can be applied to these latent variables models. Our discussion will start with a brief general description of exploratory and confirmatory factor analysis. We will then describe how the factor model parameters can be estimated using penalized estimators. Next, our discussion will turn to structural equation models, which allow for the estimation of relationships between the latent variables. We will finish the chapter by describing how regularized estimators can be applied to estimating these model parameters.

DOI: 10.1201/9780367809645-7

Factor Analysis

Factor analysis is widely used throughout the social and behavioral sciences. Researchers use this latent variable modeling approach to gain insights into a wide variety of constructs, including intelligence, personality, academic achievement and aptitude, mental illness, social phobias, and anxiety to name a few. Factor analysis can best be understood as a latent variable modeling paradigm in which a set of observed variables are the indicators of a latent variable. In this schema, the latent variable (e.g., intelligence) is of primary interest, but cannot be directly observed. However, it is theorized that the latent variable has a direct influence on each of the observed indicators (e.g., items on a scale, subscales in a battery of measures), so that they can in turn be used to gain insights into the latent variable. This idea is at the core of educational and psychological measurement of abilities.

Factor analysis is typically described as consisting of two broad types: exploratory (EFA) and confirmatory (CFA). These two types are differentiated by the degree of a priori structure that is assumed and then specified by the researcher. In the context of EFA, the researcher does not impose a specific latent structure on the observed indicators, but rather allows the optimal number of factors to be determined based on several statistical and interpretability criteria (e.g., Bandalos & Finney, 2019). This does not mean that the researcher has no preconceptions about the number or nature of the number of factors or the underlying latent structure. Indeed, it is generally best if some prior notions of the latent structure are used to assist in reaching the ultimate goal of the analysis; interpretability. However, the researcher does not explicitly link an indicator with a factor, but rather relies on the factor algorithm and set criteria to identify the optimal structure. In contrast to EFA, with CFA the researcher explicitly links the indicators with the factors to which they theoretically belong. This proposed model is then examined for its fit to the data. Typically, the fit of multiple CFA models based on differing substantive theories is compared based on a variety of statistical and theoretical criteria. The model that provides both the best statistical fit and the strongest theoretical interpretation is then selected as optimal.

Given the differences in how EFA and CFA are conceived statistically, we can consider when each is most appropriate for use. When the researcher has a definite theory regarding the latent variables and their relationships to the observed indicators, then CFA is the most appropriate technique to use. It allows for multiple models to be compared with one another and accommodates the definition of a specific structure in the data. On the other hand, the use of CFA also requires the researcher to have a strong theory that has been vetted in the literature and that has already received some empirical support, in the form of either EFA or CFA in prior research. Without such theoretical and empirical evidence, CFA may not be appropriate. Conversely, when the theory is not as strong, and/or relatively little empirical evidence exists regarding the construct, EFA may be most appropriate. EFA places a relatively light burden on the researcher regarding the expected nature of the latent constructs and their relationships with the observed indicators. At the same time, the researcher can have some general notions regarding the nature of the latent structure, including how many factors are expected and which variables should be associated with which of these factors. However, no a priori determination of these relationships is required. In the following pages, we will explore the modeling of indicators in the EFA framework using R. We also will learn how to interpret the results from such an analysis.

Common Factor Model

The factor model underlying the 4 latent variable structure is:

$$y = \Lambda\eta + \varepsilon \tag{7.1}$$

where
  y = Observed indicator variables
  η = Factors
  Λ = Factor pattern coefficients (i.e., loadings) linking observed indicators with factors
  ε = Unique variances for the indicators

In this model, we assume that ε for a given indicator variable is independent of the ε for all other indicators, and independent of η. The factor loadings reflect the relationships between the latent and observed variables, with larger values being indicative of a closer association between a latent and observed variable. The factor model in (7.1) can be used to predict the correlation (or covariance) matrix of the observed indicator variables as expressed in (7.2).

$$\Sigma = \Lambda\Psi\Lambda' + \Theta \tag{7.2}$$

where
  Σ = Model predicted correlation matrix of the indicators
  Ψ = Correlation matrix for the factors
  Θ = Diagonal matrix of unique error variances

Exploratory Factor Analysis

EFA consists of two primary steps: (1) factor extraction and (2) factor rotation, which are carried out by the software simultaneously, although the researcher must make decisions regarding the method to use for each. Factor extraction involves the initial estimation of model parameters, in particular the loadings. There are potentially as many factors as there are observed indicators in the data. Thus, for 12 items there are 12 possible factors that could be extracted. However, given that the goal of EFA is to identify a latent structure present in the data whereby a small number of latent variables account for values of the observed indicators, in practice a small number of factors will actually be retained. Below, we will demonstrate methods for determining the number of factors to retain.

Factor Extraction

A number of factor extraction methods are available, with the most popular probably being maximum likelihood (ML) and principal axis factoring (PAF). Other EFA extraction methods that are available, though used less frequently than ML and PAF, are generalized least squares, unweighted least squares, weighted least squares, alpha factoring, and image factoring, to name a few. Whichever method is used, the algorithm seeks to find estimates of factor loadings that will yield $£$ as close as possible to the observed correlation matrix, S, among the indicators. Indeed, ML extraction uses the proximity of $£$ and S to form a test statistic for evaluating the quality of a factor solution, which we will discuss later. Though ML has the advantage of providing a direct assessment of model fit, it also rests on an assumption of multivariate normality of the observed indicators. When this assumption is violated, model parameter estimates may not be accurate, and in some cases, the algorithm will not be able to find a solution (Brown, 2015; Fabrigar & Wegener, 2012). PAF does not rely on distributional assumptions about the indicators, and thus may be particularly attractive to use when the data are not normally distributed. However, it does not provide a statistical test of model fit.

Factor Rotation

With EFA, when more than one factor is retained the model identified in the extraction step is indeterminate in nature, meaning that there is an infinite number of factor loading combinations that will yield the same mathematical fit to the data; i.e., the same $£$. This leads to the question of how we determine which factor loading solutions are optimal for our purposes? This determination is made using factor rotation, which refers to the transformation of the initial set of factor loadings so as to simplify the interpretation of the results by seeking a simple structure solution. Thurstone (1947) defined simple structure as occurring when two conditions were met: (1) each factor has associated with it a subset of the indicator variables with which it is highly associated (i.e., large loadings) and (2) each indicator is highly associated with only one factor and has loadings near 0 on the other factors. Rotation adjusts all of the loadings in order to approximate this goal of simple structure without altering the underlying fit of the model. In other words, only the values of the loadings are changed in an attempt to achieve Thurstone's simple structure, in order to make interpretation of the results easier.

Factor rotation methods are generally described as being in one of two broad families, orthogonal and oblique. Orthogonal rotations constrain the correlations among factors to be 0, whereas oblique rotations allow the factors to be correlated. Within both broad rotational families there exist many varieties, differing based upon the criterion used to transform the data. As with methods of estimation, no one approach is always optimal, but perhaps the most popular orthogonal rotation method is VARIMAX, while among the oblique rotations PROMAX, and OBLIMIN are popular. The decision as to whether to use an orthogonal or oblique rotation should be based on both theoretical and empirical grounds. If the researcher anticipates that the factors will be correlated, then she should begin the analysis using an oblique rotation such as PROMAX. If the resulting correlations are small (e.g., close to 0), then the model can be refit using an orthogonal rotation. On the other hand, if the researcher feels that the correlations among factors should be constrained to 0 for some theoretical reason, then she may only use the orthogonal rotation from the beginning. However, it should be noted that if the factors are in fact correlated but an orthogonal rotation is used, the resulting factor loadings may be adversely affected, with the potential for a number of cross-loadings (variables having relatively large loadings with more than one factor) to be present.

In order to demonstrate the application of EFA, we will investigate the latent structure underlying subscores on the adult temperament scale (ATS). The proposed latent structure for these subscales appears in Table 7.1.

First, we will need to read in the data, which is available on the website for this book www.routledge.com/9780367408787. We will save it in a dataframe called example.

```
example<-read.spss("C:\\research\\regularization
book\\data\\fa_data_chapter_7.sav", to.data.frame=TRUE,
use.value.labels=FALSE)
```

Next, we will extract the eigenvalues, which can be thought of as measures of the variance accounted for by each possible latent variable underlying the observed data. Given that we have 17 observed variables in this example, there are 17 possible latent variables. Of course, we anticipate (hope) that only a few of these are truly salient and thus need to be retained. If this turns out to be true, then only a small number of the eigenvalues will be large. We

TABLE 7.1

Proposed latent structure of the Adult Temperament Scale

Subscale	Hypothesized Factor
Fear (ATS_FEAR_1)	Negative Affect
Frustration (ATS_FRUS_1)	Negative Affect
Sadness (ATS_SAD_1)	Negative Affect
Discomfort (ATS_DISC_1)	Negative Affect
Negative feelings (ATS_NEGA_1)	Negative Affect
Activation Control (ATS_ACTC_1)	Effortful Control
Attentional Control (ATS_ATTC_1)	Effortful Control
Inhibitory Control (ATS_INHC_1)	Effortful Control
Effort (ATS_EFFC_1)	Effortful Control
Sociability (ATS_SOCI_1)	Extraversion/Surgency
High Intensity Pleasure (ATS_HIP_1)	Extraversion/Surgency
Positive Affect (ATS_POSA_1)	Extraversion/Surgency
Estraversion (ATS_EXT_1)	Extraversion/Surgency
Neutral Perceptual Sensitivity (ATS_NPS_1)	Orienting Sensitivity
Affective Perceptual Sensitivity (ATS_APS_1)	Orienting Sensitivity
Associative Sensitivity (ATS_ASSO_1)	Orienting Sensitivity
Orienting (ATS_ORI_1)	Orienting Sensitivity

will extract the eigenvalues using the `eigen` function in R, and save them in a vector called `example.eigenvalues`. We can then use these to help us determine the number of factors that should be retained.

```
example.eigenvalues<-eigen(cor(example, use="complete"))
#OBTAIN EIGENVALUES#
```

One technique for using the eigenvalues to determine the number of factors to retain is the scree plot and associated tests. The subjective application of the scree plot involves plotting the eigenvalues by the factor and then looking for the point where the curve connecting these points bends. Given the subjective nature of this interpretation, researchers have developed more objective ways that the scree plot can be used to determine the number of factors to retain. We don't have space here to describe these methods in detail and refer the interested reader to Raiche, Walls, Magis, Riopel, and Blais (2012) for an in-depth description of these statistics. In R, they can be applied using the nFactors library. In particular, the `plotnScree` function produces a number of these statistics, as well as a determination of the number of factors to retain. To use this function, we use the following command sequence, which relies on the eigenvalues that we extracted above. The results, presented in the scree plot in Figure 7.1, suggest that 7 factors was the most likely number to be extracted. Given that we have 17 scale scores, it would not seem optimal for us to extract 7 factors. Note that results for one of the statistics, the Acceleration Factor, indicate that we should extract 4 factors.

```
plotnScree(nScree(x=example.eigenvalues$values, model="factors"))
#OBJECTIVE SCREE PLOT METHODS
```

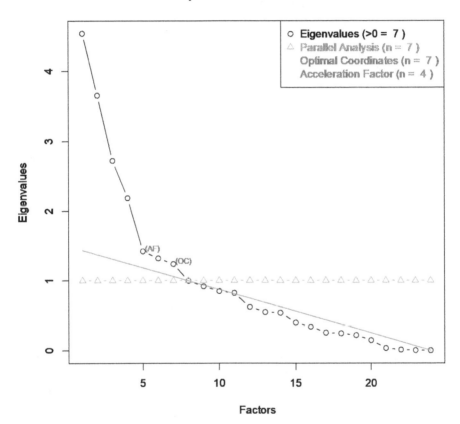

Non Graphical Solutions to Scree Test

FIGURE 7.1
Scree plot with objective statistics

Another approach for determining the number of factors to retain with an EFA is parallel analysis (PA; Horn, 1965). This method involves the random generation of a large number (e.g., 1000) of synthetic datasets based on the assumption that the null hypothesis of no underlying factors is true. The eigenvalues from these analyses on the synthetic data are then used to create distributions of eigenvalues that would be expected if no factor structure is present. The eigenvalues obtained using the observed data are then compared to these distributions in order to determine the number of factors to retain. A factor is retained if its observed eigenvalue is larger than the 95th percentile of the distribution of null factor eigenvalues generated from the random data. The synthetic data can be generated parametrically from a known distribution, such as multivariate normal with means and variances equal to the means and variances of the observed data, or it can be generated nonparametrically through the random mixing of indicator variable values within variables across observations. For example, the values of the ATS Fear subscale variable will be randomly mixed among the individuals in the sample, as will the values for Sadness, and each of the other variables. Next, the eigenvalues for this random dataset will be generated and saved, using factor analysis. This process is then repeated a large number of times. The steps underlying PA are as follows:

1. Fit an EFA to the original dataset and retain the eigenvalues for each factor.

2. Generate observed data with marginal characteristics identical to the observed data (i.e., same means and standard deviations), but with uncorrelated indicators, either by randomly sorting the values of the observed indicators or by simulating such data.

3. Fit an EFA to the generated data and retain the eigenvalues for each factor.

4. Repeat steps 2 and 3 many (e.g., 1000) times in order to develop distributions for each eigenvalue under the case where indicators are not related to one another.

5. Compare the observed eigenvalue for the first factor with the 95th percentile of the distribution of first factor eigenvalues from the generated data. If the observed value is greater than or equal to the 95th percentile value, conclude that at minimum one factor should be retained, and continue to step 6. If the observed eigenvalue is less than the 95th percentile then stop and conclude that there exist no common factors.

6. Compare the observed eigenvalue for each successive factor with the 95th percentile for the corresponding eigenvalue distribution of the generated data. If the observed eigenvalue is greater than or equal to the 95th percentile, retain that factor (e.g., the second factor, the third factor), and move to the next factor in line. This process stops when the observed eigenvalue is less than the 95th percentile of the generated data.

We can apply PA in R using the `fa.parallel` function from the `psych` library. When making the function call, we need to provide the data, request the type of factor extraction, the number of random data sets, whether there needs to be error bars, and whether the squared multiple correlations should be placed on the diagonal.

```
fa.parallel(example, fm="pa", n.iter=1000, error.
bars=FALSE, SMC=FALSE) #PARALLEL ANALYSIS#
```

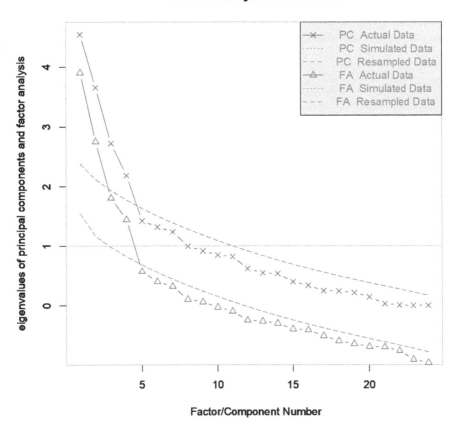

Parallel Analysis Scree Plots

FIGURE 7.2
Scree plot for parallel analysis

Focusing on the factor analysis (rather than principal components) results, we see that 4 of the observed eigenvalues exceed the 95th percentile from the synthetic dataset. Thus, we would retain 4 factors, based on the results of PA.

Another method for ascertaining the number of factors to retain that we will consider is a very simple structure (VSS). This method was originally proposed by Revelle and Rocklin (1979), and involves the examination of the ability of a factor solution to reproduce the observed variable correlation matrix by a particular factor solution, assuming that each indicator is only associated with a single latent variable; i.e., exhibits very simple structure. Recall from chapter 4 that achieving a simple structure solution is usually the ultimate goal of rotation in factor analysis. Thus, it is reasonable to ascertain how well a factor solution can reproduce the observed covariance/correlation matrix assuming that a simple structure does hold. The VSS approach uses the set of largest loadings for each variable to calculate a predicted correlation matrix, which is then used to obtain a residual correlation matrix as described earlier. In order to ascertain the closeness of the predicted to the true values, the VSS statistic is calculated as

$$VSS = 1 - \frac{SS\left(r_{residual}\right)}{SS\left(r_{observed}\right)}$$

where

$SS\left(r_{residual}\right)$ = Sum of squared residual correlations
$SS\left(r_{observed}\right)$ = Sum of squared observed correlations

VSS values are calculated for each factor solution, and the ratio of the sum of squares represents the proportion of the observed correlation matrix that is not explained by the factors. Therefore, large values of the VSS statistic indicate a relatively good fit, as most of the observed correlation is explained by the factor solution. The optimal factor solution corresponds to the maximum of VSS.

The function for conducting VSS is available in the psych library and can be carried out in R using the following command. Here we specify the dataframe, the maximum number of factors to retain, the type of rotation, should the diagonal of the covariance matrix be fit, the type of factor extraction, and whether the values should be plotted. If we want to allow no cross-loadings (the 1 line) then we would retain 4 factors, as this is where the line reaches its maximum.

```
example.vss<-vss(example, n=8, rotate="promax", diagonal=FALSE,
fm="pa", plot=TRUE, SMC=FALSE) #VSS#
```

Velicer (1976) proposed a method for determining the number of factors to retain that is based upon an examination of the average squared partial correlations among the observed indicators, accounting for the influence of the latent variables. This method, called minimum average partial (MAP), is carried out using a multiple-step procedure. In the first step, the correlations among the observed variables are calculated, squared, and then the squares are averaged. In the second step, the squared correlations among the indicators are again calculated and averaged, in this case after partialing out the first latent variable obtained using EFA. In the third step, the average squared correlation among the observed variables is again calculated, this time

Very Simple Structure

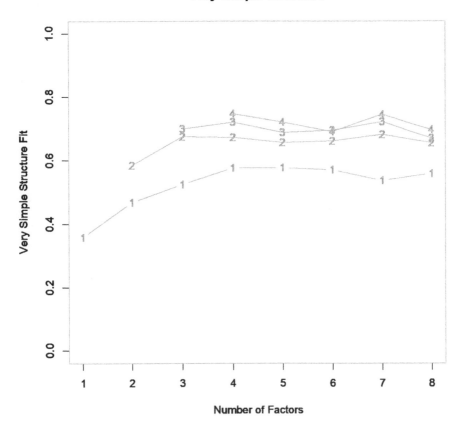

FIGURE 7.3
Very simple structure plot

partialing out the first two latent variables. These steps are repeated for the first p-1 factors, where p is the number of observed indicators. The researcher then would retain the number of factors corresponding to the minimum average squared partial correlation, as this corresponds to the point at which the maximum amount of systematic variance in the observed indicators is accounted for by the latent variables (Velicer). Simulation research has consistently demonstrated that MAP is one of the more accurate methods for determining the number of factors to retain (Caron, 2018; Ruscio & Roche, 2012; Garrido, Abad, & Ponsoda, 2011; Zwick & Velicer, 1986).

The MAP values are produced by the VSS function and retained in the map object within the output. We print them out as follows, and see that the minimum value is associated with the 5 factor solution.

```
example.vss$map #MAP#
[1] 0.05596510 0.05748171 0.05497506 0.04781389 0.04726716
0.05386272 0.05719763 0.06310526
```

One of the oldest and still most commonly used methods for determining the number of factors to retain involves an examination of the residual correlation matrix. In equation (7.2), we saw that the factor model parameters can be used to calculate the predicted correlation matrix for the observed variables. We then calculate the difference between the observed and model predicted correlations in order to retain the residual correlations. By convention (Thompson, 2004; Gorsuch, 1983), residual correlations with an absolute value greater than 0.05 are considered to be too large so that a good solution is one that produces few residual correlations greater than 0.05, in absolute value. Thus, an EFA solution that yields a large proportion of residual correlations that exceed 0.05 is not optimal. The residual correlations can be obtained easily in R, and the script to do so is available at the website for this book: www.routledge.com/9780367408787. The final step in these calculations for each factor solution from 1 to 5 appears as follows:

```
##RESIDUAL CORRELATIONS##
example.fa1.resid.large.sum/(sum(nonNA_counts1)/2)
#CALCULATE PROPORTION OF RESIDUALS THAT ARE LARGE#
[1] 0.5190311

example.fa2.resid.large.sum/(sum(nonNA_counts2)/2)
#CALCULATE PROPORTION OF RESIDUALS THAT ARE LARGE#
[1] 0.4636678

example.fa3.resid.large.sum/(sum(nonNA_counts3)/2)
#CALCULATE PROPORTION OF RESIDUALS THAT ARE LARGE#
[1] 0.3460208

example.fa4.resid.large.sum/(sum(nonNA_counts4)/2)
#CALCULATE PROPORTION OF RESIDUALS THAT ARE LARGE#
[1] 0.06920415

example.fa5.resid.large.sum/(sum(nonNA_counts5)/2)
#CALCULATE PROPORTION OF RESIDUALS THAT ARE LARGE#
[1] 0.04152249
```

For the 1, 2, and 3 factor solutions a large proportion of the residual correlations were greater than 0.05. However, for the 4 and 5 factor solutions the proportion of residual correlations was below 0.1. Furthermore, the addition of a 5th factor does not reduce the proportion of large residual correlations by much. Therefore, we would conclude that 4 factors should be retained, based on the residual correlations.

In order to determine the number of factors to retain, we will consider all of the results presented above together to come up with what may be the optimal solution. These results would seem to indicate that we should retain 4 factors in this case, which also matches the hypothesized latent structure. It is important to state again that we will also fit EFA for multiple numbers of factors. In this case, we fit from 1 to 5 factors, with the script appearing at www.routledge. com/9780367408787. Considering the conceptual fit of the items to the factors, in conjunction with the results described earlier, we would conclude that 4 factors is indeed the optimal solution for these data. The R code to fit this model, along with the resulting output appear below. Note that we specify the dataframe and variables to include in the analysis, the number of factors to retain, whether the residual correlations should be calculated, the type of rotation, whether the squared multiple correlation should be placed on the diagonal of the correlation matrix used in the analysis, and the factor extraction method. We then print the results by typing the name of the output file.

```
efa4.pa<-fa(example[,1:17], nfactors=4, residuals=TRUE,
rotate="promax", SMC=TRUE, fm="pa") #4 FACTORS#
efa4.pa

Factor Analysis using method =  pa
Call: fa(r = example[, 1:17], nfactors = 4, rotate =
"promax", residuals = TRUE,
    SMC = TRUE, fm = "pa")
Standardized loadings (pattern matrix) based upon
correlation matrix
              PA1    PA4    PA3    PA2    h2     u2  com
ATS_FEAR_1 -0.07   0.68  -0.05  -0.07  0.44   0.56  1.1
ATS_FRUS_1 -0.13   0.55  -0.25   0.07  0.33   0.67  1.6
ATS_SAD_1   0.11   0.63   0.18   0.25  0.55   0.45  1.6
ATS_DISC_1  0.26   0.51   0.18  -0.40  0.62   0.38  2.7
ATS_ACTC_1  0.04   0.21   0.83   0.12  0.74   0.26  1.2
ATS_ATTC_1 -0.02  -0.06   0.39  -0.13  0.19   0.81  1.3
ATS_INHC_1 -0.10  -0.13   0.53   0.06  0.28   0.72  1.2
ATS_SOCI_1 -0.04   0.04  -0.02   0.58  0.33   0.67  1.0
ATS_HIP_1   0.10  -0.09  -0.23   0.48  0.34   0.66  1.6
ATS_POSA_1  0.03   0.02   0.17   0.42  0.19   0.81  1.3
ATS_NPS_1   0.57   0.02   0.03   0.22  0.43   0.57  1.3
ATS_APS_1   0.75  -0.06   0.05  -0.02  0.54   0.46  1.0
ATS_ASSO_1  0.70   0.03  -0.16  -0.11  0.49   0.51  1.2
ATS_NEGA_1 -0.03   1.15   0.03  -0.02  1.31  -0.31  1.0
ATS_EFFC_1  0.00   0.00   1.08   0.06  1.14  -0.14  1.0
ATS_EXT_1  -0.06   0.03   0.00   1.17  1.35  -0.35  1.0
ATS_ORI_1   1.21  -0.15  -0.09  -0.01  1.33  -0.33  1.0

                         PA1   PA4   PA3   PA2
SS loadings             2.89  2.80  2.51  2.42
Proportion Var          0.17  0.16  0.15  0.14
Cumulative Var          0.17  0.33  0.48  0.62
Proportion Explained    0.27  0.26  0.24  0.23
Cumulative Proportion   0.27  0.54  0.77  1.00

  With factor correlations of
        PA1   PA4    PA3    PA2
PA1 1.00  0.38   0.13   0.17
PA4 0.38  1.00   0.03   0.00
```

```
PA3 0.13 0.03  1.00 -0.17
PA2 0.17 0.00 -0.17  1.00
```

```
Mean item complexity =  1.3
Test of the hypothesis that 4 factors are sufficient.
```

```
The degrees of freedom for the null model are  136  and the
objective function was  24.59 with Chi Square of  1291.2
The degrees of freedom for the model are 74  and the
objective function was  310051079
```

```
The root mean square of the residuals (RMSR) is  0.05
The df corrected root mean square of the residuals is  0.07
```

```
The harmonic number of observations is  60 with the
empirical chi square  46.91  with prob <  0.99
The total number of observations was  60  with Likelihood
Chi Square =  15450878763  with prob <  0
```

```
Tucker Lewis Index of factoring reliability =  -26060663
RMSEA index =  1865.456  and the 90 % confidence intervals
are  NA 1881.123
BIC =  15450878460
Fit based upon off diagonal values = 0.97
```

There is quite a lot of output for us to consider. First, we will examine the loadings to see which variables are associated with each of the factors. Recall that theoretically, there should be 4 factors as displayed in Table 7.1. By convention, we will consider loadings of 0.3 or larger as indicating that a variable is associated with a factor (Tabachnick & Fidell, 2019). Based on this criterion, it would appear that ATS_NPS_1, ATS_APS_1, ATS_ASSO_1, and ATS_ORI are associated with factor 1, which corresponds to the orienting sensitivity construct. The variables associated with factor 4 were ATS_FEAR_1, ATS_FRUS_1, ATS_SAD_1, ATS_DISC_1, and ATS_NEGA_1. Thus, it would seem to correspond to the negative affect latent trait. Factor 3 was associated with ATS_ATTC_1, ATS_ACTC_1, ATS_INHC_1, and ATS_EFFC_1, making it correspond to the effortful control construct. ATS_DISC_1, ATS_SOCI_1, ATS_HIP_1, ATS_POSA_1, and ATS_EXT_1 loaded on factor 2. This corresponds to what we would expect for the extraversion/surgency factor, except for the discomfort score. Therefore, with the exception of DISC, the results presented here correspond to what we anticipated based on theory.

The output from R also provides us with the proportion of variance associated with each of the factors, as well as the cumulative variance accounted for by all 4 factors together. We can see that factor 1 accounts for 17% of the variance in the observed data, followed by factor 4 with 16%, factor 3 with 15%, and factor 2 with 14%. Taken together, the factor solution accounted for 62% of the variance in the observed indicators. The correlations among the factors are next in the output. We can see that factors 1 (orienting sensitivity) and 4 (negative affect) had the highest such correlation, with the correlations between factor 4 and factors 2 (extraversion/surgency) and 3 (effortful control) being essentially 0. There was a small negative correlation between factors 2 and 3. Finally, a number of model fit indices appear at the end of the output. Research investigating the use of these statistics for assessing the fit of an EFA solution has been mixed at best (Finch, 2020; Clark & Bowles, 2018), and thus we will not interpret them here. Considering all of the results presented here, we can conclude that the hypothesized factor structure that appears in Table 7.1 is supported by the EFA.

Sparse Estimation via Nonconcave Penalized Likelihood in Factor Analysis Model (FANC)

Researchers (Hirose & Yamamoto, 2015) have proposed an alternative approach for estimating EFA models in the context of high-dimensional data. Their approach involves the use of the minimax convex penalty function (MC+) by [Zhang, 2010]. Hirose and Yamamoto showed that estimation of the EFA loadings through the maximization of the penalized log-likelihood function is possible as seen as follows:

$$(\Lambda, \Psi) = \text{argmax} \; l_p^{ort} (\Lambda, \Psi) \tag{7.3}$$

where

$$l_p^{ort} = l^{ort} (\Lambda, \Psi, \Phi) - N \sum\nolimits_{i=1}^{P} \sum\nolimits_{j=1}^{m} \rho P\left(\left|\lambda_{ij}\right|\right) \tag{7.4}$$

$l^{ort} (\Lambda, \Psi, \Phi) =$ Standard MLE factor loading estimates, error variances, and the factor covariance matrix (Φ)
$P =$ Penalty function
$\rho =$ Regularization parameter

The MC+ penalty function was selected for use with FANC because it has been shown to provide somewhat sparser and more efficient estimates than either LASSO or SCAD (Hirose & Yamamoto; Zhao & Yu, 2007; Zou, 2006). The MC+ penalty function is defined as

$$MC+ = \rho\left(\left|\theta\right| - \frac{\theta^2}{2\rho\gamma}\right) I\left(\left|\theta\right| < \rho\gamma\right) + \frac{\rho^2\gamma}{2} I\left(\left|\theta\right| \geq \rho\gamma\right) \tag{7.5}$$

where
$\theta =$ Model parameters (e.g., factor loadings, covariances)
$\gamma =$ Threshold value

An important aspect of using FANC is the selection of values for ρ and γ, which play crucial roles in yielding a sparse solution for the factor model. Based on the results of a small simulation study, Hirose and Yamamoto suggest the use of the Bayesian Information Criterion (BIC) for selecting these values. In other words, a range of γ and ρ values are assessed, and the combination that minimizes the BIC is used. This is the approach that was used in the current study.

Hirose and Yamamoto (2015) conducted a simulation study examining the performance of the FANC estimator and found that MC+ yielded lower MSE and higher TPR than did the LASSO, leading the researchers to suggest that the FANC algorithm should be considered for use with high-dimensional data. The FANC estimator can be applied using the fanc function from the fanc R library. Following is the function call for the 4 factor solution.

```
fanc.efa4<-fanc(as.matrix(example[,1:17]), 4, cor.
factor=TRUE, type="MC", control=list(length.rho=10, length.
gamma=10, start="cold"))
```

The results are saved in an output object called fanc.efa4. The dataframe needs to be coerced into a matrix, after which the number of factors to be retained is specified. Next, we specify that the factors should be allowed to correlate with one another and that the MC penalty should be used. In the control section, we indicate that there should be 10 different values of rho and 10 different values of gamma, leading to 100 different regularization conditions. Finally, we indicate that no starting values are given (start="cold"). We then select the optimal penalized solution based on minimizing the BIC using the following:

```
fanc.efa4.results.bic<-select(fanc.efa4,
criterion=c("BIC"))
```

We can then print out these optimal results by simply typing the name of the output file. Note that we can select the optimal model using other information indices such as the AIC, sample size adjusted BIC, or the CAIC.

```
fanc.efa4.results.bic
$loadings
17 x 4 sparse Matrix of class "dgCMatrix"
                  Factor1      Factor2      Factor3     Factor4
ATS_FEAR_1   .            .            .           0.7326833
ATS_FRUS_1  -0.17398899   .            .           0.5899394
ATS_SAD_1    .            .            0.1553380   0.7575698
ATS_DISC_1   0.03494698   0.1058059   -0.3794710   0.6358787
ATS_ACTC_1   0.81740469   .            .           0.1498382
ATS_ATTC_1   0.52803639   .            .           .
ATS_INHC_1   0.61076512   .            .           .
ATS_SOCI_1   .            .            0.6973214   .
ATS_HIP_1    .            .            0.6744970   .
ATS_POSA_1   .            .            0.4824782   .
ATS_NPS_1    .            0.7319788    .           .
ATS_APS_1    .            0.8035531    .           .
```

```
ATS_ASSO_1   .              0.7716890   .           .
ATS_NEGA_1   .              .           .           0.9797133
ATS_EFFC_1   0.97460034 .               .           .
ATS_EXT_1    .              .           0.9802169 .
ATS_ORI_1    .              0.9814061   .           .

$uniquenesses
ATS_FEAR_1 ATS_FRUS_1   ATS_SAD_1 ATS_DISC_1 ATS_ACTC_1 ATS_
ATTC_1 ATS_INHC_1 ATS_SOCI_1
  0.4434490   0.5861099   0.3628453   0.3370233   0.2378165
0.7079060   0.6092100   0.4964239
 ATS_HIP_1 ATS_POSA_1   ATS_NPS_1   ATS_APS_1 ATS_ASSO_1 ATS_
NEGA_1 ATS_EFFC_1   ATS_EXT_1
  0.5288500   0.7589237   0.4464288   0.3328774   0.3847366
0.0050000   0.0050000   0.0050000
 ATS_ORI_1
  0.0050000

$Phi
              Factor1     Factor2     Factor3     Factor4
Factor1   1.00000000 0.09556205 -0.13558140   0.08615961
Factor2   0.09556205 1.00000000   0.17774993   0.32136360
Factor3  -0.13558140 0.17774993   1.00000000 -0.03044023
Factor4   0.08615961 0.32136360 -0.03044023   1.00000000

$df
[1] 46

$BIC
[1] 585.7864

$goodness.of.fit
      GFI        AGFI        CFI       RMSEA        SRMR
0.4549553 0.2206371 0.3975281 0.5407687 0.1056426

$rho
[1] 0.3716626

$gamma
[1] 1.01
```

Using the same criterion for interpretation that was employed with the PAF estimates, we can gain insights into the latent structure as suggested by FANC. Factor 1 appears to correspond to the effortful control latent variable, factor 2 to orienting sensitivity, factor 3 to extraversion/surgency, and factor 4 to negative affect. Note that most of the loadings are smaller than those from the standard EFA, but in other respects correspond fairly closely. The correlations among the factors appear in the portion of the output labeled Phi. As with the factor loadings, these results for the correlations are similar to those from the standard EFA. As we mentioned earlier in this section, the researcher will want to examine a variety of solutions, changing both the number of factors and the regularization parameter values. From a statistical perspective, the optimal solution corresponds to the minimum information index value and may vary depending upon which is used. And of course, the selected model must be theoretically defensible, as was the case in this example.

Confirmatory Factor Analysis

We presented the common factor model in equation (7.1), which explicitly shows the link between the observed indicators (y) and the latent variables (η). This model applies to both EFA and CFA, as does the relationship between the observed variable covariance matrix and the latent structure, as described in equation (7.2). For the purposes of CFA, we would make one addition to (7.1) by including an intercept term (τ), as shown in (7.6).

$$y = \tau + \Lambda\eta + \varepsilon \qquad (7.6)$$

For EFA models, τ is generally assumed to be 0, but with CFA this is not necessarily the case. Indeed, if we are interested in comparing latent variable means across groups, the model intercepts play an important role. As was true with EFA, in CFA each of these model parameters will be estimated using the variances and covariances of the observed variables. However, whereas in EFA the focus was on deciding how many factors to extract and which type of rotation to use, with CFA these issues are not a concern. We enter a CFA analysis with a predetermined factor structure in mind thereby obviating the need to worry about determining the number of factors or identifying the maximally interpretable solution using rotation. However, new considerations do emerge regarding how the model parameters should be estimated, and which of several proposed models fits the data best. It is with these issues that we will concern ourselves next, before learning how to fit CFA models using R. We begin with model parameter estimation.

Model Parameter Estimation

There are a number of potential methods for estimating the parameters in model (7.6). Certainly the most common of these is Maximum Likelihood (ML), which was discussed briefly in the previous chapter. ML estimates model parameters (i.e., factor loadings, variances, covariances, intercepts, and error variances) that minimize the criterion in equation (7.7).

$$F_{ML} = \frac{1}{2} tr \left[\left([S - \Sigma] \Sigma^{-1} \right)^2 \right] \tag{7.7}$$

where
  S = Covariance matrix among the observed indicators
  Σ = Model predicted covariance matrix among the observed indicators

As we noted in Chapter 2, the ML algorithm relies on an assumption that the observed indicators are multivariate normal. If they are not, the estimated standard errors will be incorrect, leading to improper significance test results (Yuan, Bentler, & Zhang, 2005).

One alternative method of parameter estimation that has been shown to be effective when the indicators are not multivariate normal is weighted least squares (WLS), which has the following fit criterion:

$$F_{WLS} = (S - \Sigma)' W^{-1} (S - \Sigma) \tag{7.8}$$

In (7.8) the weight matrix, W, is the asymptotic covariance matrix of the elements contained in the observed sample covariance matrix. The logic behind using this particular weight matrix is that those elements (i.e., observed variances and covariances) that have less sampling variability (i.e., smaller values in W) will receive greater weight than those elements that have greater sampling variability. A potential problem with WLS is that its computational complexity renders it less than optimal when the sample size is not large (Finney & Distefano, 2013). A recommended approach for addressing this problem involves the use of the diagonal of W in (3.3), rather than the entire matrix. More specifically, the fit criterion for this diagonally weighted least squares approach (DWLS) takes the form:

$$F_{DWLS} = (S - \Sigma)' (diag W)^{-1} (S - \Sigma) \tag{7.9}$$

Thus, the residuals of the observed covariance matrix are weighted not by the full covariance matrix of S, but rather only by the variances of the elements in S. To avoid the bias that would result in estimating model parameter standard errors and the model chi-square test without considering the full covariance matrix, information from W is incorporated into their estimation without the use of the full WLS approach. The reader interested in the more technical issues of this approach is referred to Wirth and Edwards (2007), Flora and Curran (2004), and Muthén, du Toit, and Spisic (1997).

The methods for estimation mentioned earlier only scratch the surface of the options available for estimating model parameters. These were selected because they are (a) effective, (b) widely used, and (c) available in most standard SEM software packages. At the same time, we also acknowledge that there do exist a wide array of model parameter estimation techniques. The reader will want to immerse themselves in the literature on this topic in order to gain a full understanding of the options. For this purpose, we recommend Finney and DiStefano (2013), Brown (2015), and Kline (2016). We would also add that based on our experience and available research in the area, the majority of research scenarios involving SEM can be addressed using one of these estimation methods earlier.

Assessing Model Fit

Once parameters have been estimated, the researcher must ascertain whether the model fits the data. Put more simply, is the model able to reproduce with accuracy the covariance matrix of the observed variables? Many ways exist to assess model fit, and we will only discuss those that have been shown to be most accurate, and that are widely available in computer software. Perhaps the most common method is the chi-square goodness-of-fit test. However, this statistic is not particularly useful in practice because it tests the null hypothesis that $\Sigma = S$, which is very restrictive. The test will almost certainly be rejected when the sample size is sufficiently large (Bollen, 1989, 1990). In addition, the chi-square test relies on the assumption of multivariate normality of the indicators, which may not be tenable in many situations. As to this latter point, several corrections to the chi-square value have been suggested for cases where data are not multivariate normal and bias in the statistic is feared. One of these is the Satorra-Bentler correction (Satorra & Bentler, 1994), which adjusts the standard ML-based statistic as follows:

$$\chi^2_{SB} = d^{-1} \chi^2_{ML} \tag{7.10}$$

where

χ^2_{ML} = Standard chi-square statistic from ML

d = Scaling factor associated with the multivariate kurtosis in the observed data

When kurtosis of the observed indicators differs, the χ^2_{SB} test has been found to be somewhat biased (e.g., Curran, West, & Finch, 1996). To correct for this bias, Yuan, Bentler, and Kano (1997) offered a test statistic (χ^2_{YB}) that used a slightly different scaling value and that appears to outperform χ^2_{SB} in such cases. A third alternative to the standard χ^2_{ML} test statistic for use when the observed data are not multivariate normal is based on work by Bollen and Stine (1992), who proposed using a bootstrap approach to calculating estimates of the chi-square test statistic, as well as adjustments to the model standard errors when the data are not normally distributed. The bootstrap (Efron, 1982), which involves repeatedly resampling from the original sample with replacement, has a long history in statistics and is widely used in a number of applications. In the context of CFA, Bollen and Stine described a method involving the bootstrap for developing the distribution of the chi-square statistic under the null hypothesis and the corresponding development of a hypothesis test that would be robust to departures from normality. In particular, the χ^2_{ML} for the original data is calculated, and then the observed data are transformed to match the covariance matrix implied by the CFA model. This transformed dataset is then resampled using the bootstrap many times (e.g., 1000) and for each of these samples, the χ^2_{ML} is calculated and saved, creating a distribution of χ^2_{ML} when the null hypothesis of good model fit is true. The original χ^2_{ML} is then compared to this distribution, and the *p*-value for the hypothesis test is the proportion of bootstrap chi-square values that exceed the χ^2_{ML} from the original dataset.

Whereas the chi-square statistics described earlier provide a direct test of the null hypothesis of exact model fit, researchers using CFA typically refer to a wide array of relative fit indices as well. Indeed, these may prove to be more useful in assessing the results of a CFA compared to the chi-square test, for reasons described previously. When using such fit indices, researchers are encouraged to refer to several and consider the collective results rather than taking each as an absolute up or down vote on the fit of the model. In other words, taken together what do the indices suggest about model fit? Among the most popular of such indices is the Root Mean Square Error of Approximation (RMSEA), which is calculated as:

$$RMSEA = \sqrt{\frac{\chi^2_T - df_T}{df_T(n-1)}} \qquad (7.11)$$

where

χ^2_T = ML-based chi-square test for the target model; i.e., the model of interest

df_T = Degrees of freedom for the target model (number of observed covariances and variances minus number of parameters to be estimated)

n = Sample size

By convention, values of RMSEA $\leq$ 0.05 are taken to indicate good model fit, and values between 0.05 and 0.08 are seen as indicative of adequate model fit (Kline, 2016). When RMSEA exceeds 0.08, the fit is said to be poor.

A second popular and proven fit statistic is the Comparative Fit Index (CFI):

$$CFI = 1 - \frac{Max(\chi^2_T - df_T, 0)}{Max(\chi^2_0 - df_0, 0)} \qquad (7.12)$$

where

χ^2_0 = ML-based chi-square test for the null model in which no relationships between the latent and observed variables are hypothesized to exist

df_0 = Degrees of freedom for the null model

A closely related fit statistic to the CFI is the Tucker–Lewis Index (TLI), sometimes also referred to as the non-normed fit index (NNFI). It is calculated as:

$$TLI = \frac{\dfrac{\chi_0^2}{df_0} - \dfrac{\chi_T^2}{df_T}}{\dfrac{\chi_0^2}{df_0} - 1} \tag{7.13}$$

One recommendation for when a model is considered to exhibit good fit is when values of CFI and TLI are 0.95 or higher (Hu & Bentler, 1999). Of course, as with any type of descriptive fit value, other recommendations might be found to support a different cutoff (e.g., 0.90), and no single cut-off should be taken as absolutely accurate. Rather, higher values on these indices indicate a better fit of the model to the data.

A fourth index of model fit that is frequently used is the Standardized Root Mean Square Residual (SRMR). SRMR is calculated as:

$$SRMR = \frac{\sum \left(r_{observed\,j,k} - r_{predicted\,j,k} \right)^2}{\left(p(p+1)/2 \right)} \tag{7.14}$$

where

$r_{observed\,j,k}$ = Observed correlation between indicator variables j and k

$r_{predicted\,j,k}$ = CFA model predicted correlation between indicator variables j and k

p = Number of observed indicator variables

Hu and Bentler (1999) suggested values of SRMR ≤ 0.08 suggest a good model fit to the data.

When our interest is in comparing the fit of two or more models, we have several statistical options for making the determination. When two models are nested within one another (i.e., one model is a more constrained version of another), we can use their individual chi-square goodness of fit test statistics to create a test comparing their relative fit. This test is possible because the difference in two chi-square values is itself a chi-square, with degrees of freedom equal to the difference in the degrees of freedom for each statistic. We can calculate this statistic as:

$$\chi_C^2 - \chi_U^2 = \chi_\Delta^2 \tag{7.15}$$

With degrees of freedom equal to

$$df_C - df_U = df_\Delta \tag{7.16}$$

where
   χ_C^2 = Chi-square statistic from constrained model
   χ_U^2 = Chi-square statistic from unconstrained model
   df_C = Degrees of freedom for constrained model
   df_U = Degrees of freedom for unconstrained model

When χ_Δ^2 is statistically significant at the desired level of α (e.g., 0.05), we can conclude that the relative fit of the two models is statistically different; i.e., one fits better than the other.

Another statistical approach for comparing model fit involves the use of information indices, which are simply measures of variance not explained by the model, with an added penalty for model complexity. Among the most popular of these indices are the Akaike Information Criterion (AIC; Akaike, 1973), the Bayesian Information Criterion (BIC; Schwarz, 1978), and the sample size adjusted BIC (SBIC; Tofighi & Enders, 2007). Each of these statistics is based upon the model chi-square and is interpreted such that the model with the lower value exhibits a better fit to the data. An advantage of these information indices over the chi-square difference test is that models do not need to be nested for comparisons to be conducted. The information indices are calculated as follows:

$$AIC = \chi_M^2 + 2q \tag{7.17}$$

$$BIC = \chi_M^2 + qln(n)v \tag{7.18}$$

$$SBIC = \chi_M^2 + ln\left[(n+2)/24\right]\left[\frac{v(v+1)}{2} - df\right] \tag{7.19}$$

where
   χ_M^2 = Model chi-square value
   q = Number of parameters estimated in the model
   v = Number of observed variables
   n = Sample size
   df = Model degrees of freedom

We do remind the reader that it is important to not only focus on fit statistics and indices when deciding if a model fits the data. If we maintain fit index tunnel vision we may miss other important information that could indicate a problem with the fit of the model. The analyst will want to, for example, inspect the parameter estimates to be certain the values are with the appropriate range as well as inspect other information, such as the residuals, as was discussed in with EFA. The examination of all the information will help to ensure the fit is fully evaluated.

Now that we have reviewed the theory underlying CFA, let's see how to fit these models using R. There are multiple ways to do this, with one of the most flexible approaches contained in the `lavaan` package. Following are the commands for fitting a CFA model for the ATS, which we described earlier in the chapter. First, we must specify the model and save it in a model object. Because we are fitting a CFA model, each indicator is explicitly assumed to be linked to its hypothesized factor. In addition, rather than identifying the model by setting a referent indicator factor loading to 1, we will estimate all of the loadings (specified by placing the NA* in front of the lead variable for each indicator). Model identification is achieved by setting each factor variance to 1, which is done with the `negative~~1*negative` command for each factor. Once we specify the model, we then fit it using the `cfa` function and then summarize the results. We need to explicitly request both the measures of model fit and the standardized estimates.

```
###CONFIRMATORY FACTOR ANALYSIS###
example.cfa.model <- '
        negative =~ NA*ATS_FEAR_1 + ATS_FRUS_1 + ATS_SAD_1 +
           ATS_DISC_1
        control =~  NA*ATS_INHC_1  + ATS_ATTC_1
        extraversion =~ NA*ATS_SOCI_1 + ATS_HIP_1 + ATS_POSA_1
        sensitivity =~ NA*ATS_NPS_1 + ATS_APS_1 + ATS_ASSO_1
        negative~~1*negative
        control~~1*control
        extraversion~~1*extraversion
        sensitivity~~1*sensitivity'
example.cfa.results <- cfa(example.cfa.model, data=example)

summary(example.cfa.results, fit.measures=T, standardized=T)
```

The results of our analysis appear below. We will only highlight certain portions, and the interested reader is encouraged to learn more from the lavaan tutorial available at https://lavaan.ugent.be/tutorial/index.html.

```
lavaan 0.6-6 ended normally after 59 iterations

    Estimator                                         ML
    Optimization method                           NLMINB
    Number of free parameters                         30
    Number of observations                            60
Model Test User Model:
    Test statistic                                67.115
    Degrees of freedom                                48
    P-value (Chi-square)                           0.036

Model Test Baseline Model:

    Test statistic                               162.693
    Degrees of freedom                                66
    P-value                                        0.000

User Model versus Baseline Model:

    Comparative Fit Index (CFI)                    0.802
    Tucker-Lewis Index (TLI)                       0.728
```

```
Loglikelihood and Information Criteria:

  Loglikelihood user model (H0)                  -917.617
  Loglikelihood unrestricted model (H1)          -884.059
  Akaike (AIC)                                   1895.234
  Bayesian (BIC)                                 1958.064
  Sample-size adjusted Bayesian (BIC)            1863.706

Root Mean Square Error of Approximation:

  RMSEA                                             0.081
  90 Percent confidence interval - lower           0.022
  90 Percent confidence interval - upper           0.125
  P-value RMSEA <= 0.05                             0.144

Standardized Root Mean Square Residual:

  SRMR                                              0.114

Parameter Estimates:

  Standard errors                              Standard
  Information                                  Expected
  Information saturated (h1) model            Structured

Latent Variables:
                  Estimate  Std.Err  z-value  P(>|z|)   Std.lv  Std.all
  negative =~
    ATS_FEAR_1      0.293    0.112    2.609    0.009     0.293    0.339
    ATS_FRUS_1      0.030    0.080    0.375    0.708     0.030    0.035
    ATS_SAD_1       0.306    0.129    2.373    0.018     0.306    0.292
    ATS_DISC_1      1.264    0.230    5.488    0.000     1.264    1.259
  control =~
    ATS_INHC_1      0.091    0.154    0.589    0.556     0.091    0.116
    ATS_ATTC_1      0.434    0.567    0.764    0.445     0.434    0.493
  extraversion =~
    ATS_SOCI_1      0.316    0.142    2.220    0.026     0.316    0.320
    ATS_HIP_1       0.513    0.167    3.066    0.002     0.513    0.523
    ATS_POSA_1      0.175    0.102    1.716    0.086     0.175    0.228
  sensitivity =~
    ATS_NPS_1       0.442    0.130    3.399    0.001     0.442    0.466
    ATS_APS_1       0.919    0.144    6.365    0.000     0.919    0.858
    ATS_ASSO_1      0.570    0.131    4.340    0.000     0.570    0.584

Covariances:
                  Estimate  Std.Err  z-value  P(>|z|)   Std.lv  Std.all
  negative ~~
    control         0.175    0.293    0.597    0.551     0.175    0.175
    extraversion   -0.626    0.222   -2.814    0.005    -0.626   -0.626
    sensitivity     0.346    0.124    2.793    0.005     0.346    0.346
  control ~~
    extraversion   -0.935    1.224   -0.763    0.445    -0.935   -0.935
    sensitivity    -0.118    0.325   -0.365    0.715    -0.118   -0.118
  extraversion ~~
    sensitivity     0.298    0.245    1.216    0.224     0.298    0.298
```

```
Variances:
```

	Estimate	Std.Err	z-value	P(>|z|)	Std.lv	Std.all
negative	1.000				1.000	1.000
control	1.000				1.000	1.000
extraversion	1.000				1.000	1.000
sensitivity	1.000				1.000	1.000
.ATS_FEAR_1	0.663	0.122	5.423	0.000	0.663	0.885
.ATS_FRUS_1	0.722	0.132	5.480	0.000	0.722	0.999
.ATS_SAD_1	1.003	0.182	5.504	0.000	1.003	0.915
.ATS_DISC_1	-0.590	0.573	-1.030	0.303	-0.590	-0.585
.ATS_INHC_1	0.599	0.111	5.378	0.000	0.599	0.987
.ATS_ATTC_1	0.585	0.495	1.181	0.238	0.585	0.757
.ATS_SOCI_1	0.876	0.166	5.283	0.000	0.876	0.898
.ATS_HIP_1	0.699	0.177	3.954	0.000	0.699	0.726
.ATS_POSA_1	0.560	0.103	5.450	0.000	0.560	0.948
.ATS_NPS_1	0.703	0.139	5.068	0.000	0.703	0.783
.ATS_APS_1	0.304	0.181	1.682	0.093	0.304	0.265
.ATS_ASSO_1	0.629	0.136	4.624	0.000	0.629	0.659

The fit statistics suggest that the model does not fit the data very well, given that CFI (0.802), TLI (0.728), RMSEA (0.081), and the SRMR (0.114) are all outside of the acceptable bounds cited earlier. The factor loadings also reflect poor model fit for a number of the observed indicators, with standardized values below 0.3. Finally, the correlations among the factors show some very strong negative relationships involving extraversion.

In practice, we would want to investigate a variety of alternative estimators and factor structures before coming to a final decision regarding the optimal model. However, given that our focus in this book is on regularized estimators, we will turn our attention to several approaches to fitting penalized latent variable models using R. The first of these is regularized SEM (RegSEM).

RegSEM

Jacobucci, Grimm, and Mcardle (2016) described the use of regularization techniques for full information estimation SEM (RegSEM) as a means to identify models that are simpler and more generalizable to the population. Because model parameters, such as factor loadings, are allowed to vary during estimation even as the regularization methodology works to keep the model simple (minimize the number of non-zero parameters), it is better able to identify model misspecifications such as cross-loadings than is the standard MLE approach. Jacobucci et al. argue that applied researchers may have more flexibility with direct penalties in the estimation model to prevent overfitting of structural parameters, while at the same time not unnecessarily constraining model parameters to be a specific value (i.e., 0). The fit function for RegSEM is based upon that of MLE, with the addition of a penalty function, as seen in the following equation:

$$F_{regsem} = F_{ML} + \lambda P(\theta) \tag{7.20}$$

where

F_{ML} = The standard MLE fit function
λ = A tuning parameter value between zero and infinity

$P(\cdot)$ = A penalty function which is the sum of one or more of the model's parameter matrices. The larger the value of λ in equation (7.20), the greater the penalty that is placed on the model parameter estimates, which will lead to their shrinkage. Thus, estimates that would be relatively close to zero based upon MLE will be zero with RegSEM. On the other hand, for parameters that are not near zero, the estimates obtained from RegSEM, while shrunken, will not be zero. Thus, non-trivial cross-loadings that would be set to zero in the context of MLE would be estimated as greater than zero by RegSEM. Likewise, paths between latent variables that are not expected, and thus set to zero in MLE would also be estimated by RegSEM, thereby avoiding the misspecification problem in the structural portion of the model. In either case, the estimation bias associated with misspecified models can be greatly ameliorated by RegSEM, as demonstrated in prior work (Jacobucci et al., 2016).

RegSEM can accommodate the Ridge penalty RegSEM (Ridge), as well as the Lasso penalty (RegSem Lasso). The penalty term for the Ridge fitting function in the context of SEM is

$$P(\cdot)_{ridge} = \|\theta\|_2 == \Sigma_{j=1}^{p} \theta_j^2 \tag{7.21}$$

where

θ_j = Parameter j.

The lasso penalty for RegSEM is similar in form:

$$P(\cdot)_{lasso} = \|\theta\|_1 = \Sigma_{j=1}^{p} |\theta_j| \tag{7.22}$$

In this manuscript, the RegSEM estimators are referred to as ML-Lasso (regsem) and ML-Ridge (regsem). Collectively, they will be referred to as RegSEM.

A key aspect of successfully using any regularization method is determining the optimal value for the tuning parameter λ. The selection of λ for a given problem is determined by fitting separate models for a large number of different values (e.g., 20–100; Jacobucci et al., 2016), ranging from 0 (no penalty) to a value of λ that results in estimation problems (Jacobucci et al., 2016), at which point the model with the optimal fit, as measured by either the root-mean-square error of approximation (RMSEA) or the Bayesian Information Criterion (BIC), is selected as being optimal. An example of estimation problems manifesting themselves in RegSEM would be when a latent variable is no longer present in the model due to all of its loadings being shrunken to zero. A full discussion of the details of these shrinkage methods is beyond the purview of the current paper. Interested readers are referred to Jacobucci et al. (2016) and Hastie et al. (2015), where these estimators are discussed in great detail.

We can fit the RegSEM estimator using functions from the regsem library. First, we must determine the optimal value of λ, which is done using cross-validation just as we described in Chapter 2. We will save the results for use with other functions later. For the cv.regsem function, we must specify the results from the CFA model that we fit using lavaan. We then must specify which model parameters we want to penalize (loadings in this case), as well as the type of model function, the number of lambda values to assess, and the difference between the lambda values that we want to assess (0.01 in this example). We can then use the summary function to see the results, including the optimal value of λ. For this example, the optimal tuning parameter value is 0.61.

```
#FIT LASSO CFA#
lasso.cv.results <- cv_regsem(example.cfa.
results,type="lasso", pars_pen=c("loadings"),
gradFun="ram",n.lambda=100, jump=0.01)

summary(lasso.cv.results)
Number of parameters regularized: 8
Lambda ranging from 0 to 0.69
Lowest Fit Lambda: 0.61
Metric: BIC
Number Converged: 39
```

We can plot the factor loading estimates by the tuning parameter value, basing the optimal value on minimizing BIC.

```
plot(lasso.cv.results, show.minimum="BIC")
```

In order to see the factor loading estimates for the optimal solution, we can type the following command:

```
lasso.cv.results$final_pars
```

negative -> ATS_FEAR_1	negative -> ATS_FRUS_1	negative -> ATS_SAD_1
0.260	0.000	0.247
negative -> ATS_DISC_1	control -> ATS_INHC_1	control -> ATS_ATTC_1
1.053	0.000	0.000
extraversion -> ATS_SOCI_1	extraversion -> ATS_HIP_1	extraversion -> ATS_POSA_1

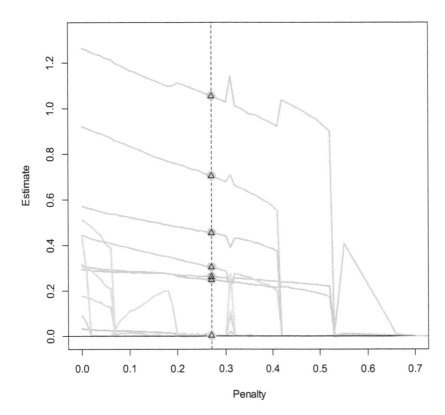

FIGURE 7.4
Factor loading estimates by penalty parameter value

0.000	0.000	0.000
sensitivity -> ATS_NPS_1	sensitivity -> ATS_APS_1	sensitivity -> ATS_ASSO_1
0.302	0.702	0.453
ATS_FEAR_1 ~~ ATS_FEAR_1	ATS_FRUS_1 ~~ ATS_FRUS_1	ATS_SAD_1 ~~ ATS_SAD_1
0.621	0.723	0.956
ATS_DISC_1 ~~ ATS_DISC_1	ATS_INHC_1 ~~ ATS_INHC_1	ATS_ATTC_1 ~~ ATS_ATTC_1
-0.302	0.607	0.773
ATS_SOCI_1 ~~ ATS_SOCI_1	ATS_HIP_1 ~~ ATS_HIP_1	ATS_POSA_1 ~~ ATS_POSA_1
0.973	0.936	0.588
ATS_NPS_1 ~~ ATS_NPS_1	ATS_APS_1 ~~ ATS_APS_1	ATS_ASSO_1 ~~ ATS_ASSO_1
0.726	0.455	0.625
negative ~~ control	negative ~~ extraversion	negative ~~ sensitivity
-284.005	-2026.294	0.384
control ~~ extraversion	control ~~ sensitivity	extraversion ~~ sensitivity
-393.766	115.125	1191.334

Using RegSEM, we can fit the ridge estimator in much the same way that we did the lasso. We simply indicate type="ridge". Otherwise, the command sequence matches what we did for the lasso estimator.

```
#FIT RIDGE CFA#
ridge.cv.results <- cv_regsem(example.cfa.
results,type="ridge", pars_pen=c("loadings"),
gradFun="ram",n.lambda=100, jump=0.01)

plot(ridge.cv.results, show.minimum="BIC")
```

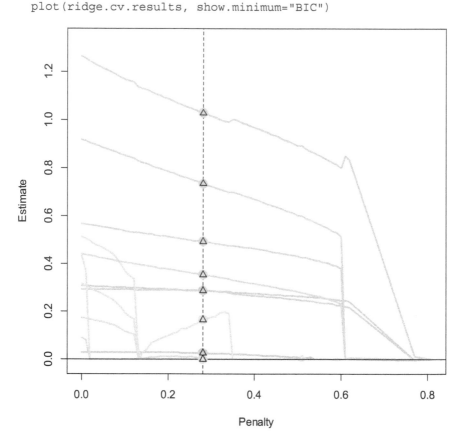

FIGURE 7.5
Factor loading estimates by penalty parameter value

```
summary(lasso.cv.results)

CV regsem Object
 Number of parameters regularized: 12
 Lambda ranging from 0 to 0.7
 Lowest Fit Lambda: 0.27
 Metric: BIC
 Number Converged: 57

ridge.cv.results$final_pars
       negative -> ATS_FEAR_1      negative -> ATS_FRUS_1      negative -> ATS_SAD_1
                   0.260                       0.000                       0.247
       negative -> ATS_DISC_1      control -> ATS_INHC_1       control -> ATS_ATTC_1
```

1.053	0.000	0.000
extraversion -> ATS_SOCI_1	extraversion -> ATS_HIP_1	extraversion -> ATS_POSA_1
0.000	0.000	0.000
sensitivity -> ATS_NPS_1	sensitivity -> ATS_APS_1	sensitivity -> ATS_ASSO_1
0.302	0.702	0.453
ATS_FEAR_1 ~~ ATS_FEAR_1	ATS_FRUS_1 ~~ ATS_FRUS_1	ATS_SAD_1 ~~ ATS_SAD_1
0.621	0.723	0.956
ATS_DISC_1 ~~ ATS_DISC_1	ATS_INHC_1 ~~ ATS_INHC_1	ATS_ATTC_1 ~~ ATS_ATTC_1
-0.302	0.607	0.773
ATS_SOCI_1 ~~ ATS_SOCI_1	ATS_HIP_1 ~~ ATS_HIP_1	ATS_POSA_1 ~~ ATS_POSA_1
0.973	0.936	0.588
ATS_NPS_1 ~~ ATS_NPS_1	ATS_APS_1 ~~ ATS_APS_1	ATS_ASSO_1 ~~ ATS_ASSO_1
0.726	0.455	0.625
negative ~~ control	negative ~~ extraversion	negative ~~ sensitivity
-284.005	-2026.294	0.384
control ~~ extraversion	control ~~ sensitivity	extraversion ~~ sensitivity
-393.766	115.125	1191.334

The results for the two approaches are quite similar to one another.

Penfa

An alternative approach to regularization in the context of factor analysis has been proposed by Geminiani, Marra, and Moustaki (2021). This methodology is based upon a trust region approach to estimating model parameters. Geminiani pointed out that a problem with the standard approaches to applying regularization to FA models is that all estimates are penalized equally, and are thus not sensitive to potential differences in parameters. As a consequence, they may yield either overly complex (too little regularization across parameters) or overly shrunken (too much regularization across parameters). Thus, the regularization framework proposed by Geminiani et al. allows for specific regularization applied to each model parameter, in contrast to other approaches applied to FA, which use a single penalty across all of the parameters that are to be regularized. As an example, if we use the penfa technique to regularize factor loadings, there will be unique shrinkage applied to each of the loadings, whereas other approaches apply a single shrinkage to the entire set of loadings as a unit. Readers interested in more technical details of the method are encouraged to read Geminiani et al.

In order to apply the penfa method to our example, we will need to load the penfa library from R. With this library, we have the options of the lasso, ridge, adaptive lasso, mcp, and scad penalty functions. First, let's apply the lasso penalty to the example data that we have been working with. We will save the output in an object called penfa.lasso. The primary function call is penfa, and requires us to give the model, which can be a lavaan model object.

We must specify the dataset, the information function to be used (fisher or hessian), and the penalty to be applied (lasso, ridge, alasso, mcp, or scad). In addition, we have the option of conducting multiple groups CFA, while applying a regularization approach to group differences on one or more model parameters. In this case, we do not conduct multiple groups CFA and thus don't apply regularization to group differences. Next, we must specify the model parameters to be regularized, which are specified in the eta= subcommand. In this example, we request shrinkage for the factor loadings (lambda) and indicate that the starting value for the penalty tuning parameter starts at 0.1. The next line in the function call indicates that no parameters are to be regularized for group differences. There are two possible strategies for determining the optimal penalty parameter, fixed and auto. In this case, we use the auto approach, allowing the algorithm to guide its own search, beginning at the starting point that we provided in the eta= subcommand. The alternative would be to use a set of possible tuning parameters, in which case we would use the fixed option for the strategy. Unless we have predetermined values for the tuning parameters, the auto option will generally be preferable. Once the algorithm has completed, we can use the summary command to obtain the output.

```
library(penfa)
##LASSO PENALTY##
penfa.lasso <- penfa(model = example.cfa.model,
data = example,
information = "fisher",
pen.shrink = "lasso",
pen.diff = "none",
eta = list(shrink = c("lambda" = 0.1),
diff = c("none" = 0)),
strategy = "auto",
verbose = FALSE)

summary(penfa.lasso)
```

The output appears below and has a very similar format to that we obtain from lavaan. The output tells us that the factor loadings are penalized and that the factor covariances, variances, and error variances are either fixed or freely estimated. The heading of the output includes information about the sample size, the number of groups, the number of latent and observed variables, and the estimator that was used. In addition, the software provides us with two information indices that can be used to compare model fit, the generalized information criterion (GIC) and the generalized Bayesian information criterion (GBIC). These correspond closely to the commonly used AIC and BIC, which we've used throughout the book. And as with those other information criteria, smaller values indicate a better fitting model.

With respect to the loadings, notice that those for indicators associated with the control and extraversion factors are either 0 or very close to 0. In other words, when we apply the lasso penalty there is very little evidence for the existence of these factors, which is contrary to our theoretical expectations. In contrast, multiple loadings for the other factors were statistically significantly different from 0 (95% confidence interval did not include 0). These results are very similar to those obtained from regsem, and quite different from those produced by lavaan, particularly with respect to the control factor. With respect to the indicator error variances, the penfa and lavaan approaches yielded relatively similar results.

```
penfa 0.0.0.9000 reached convergence

    Number of observations                              60
    Number of groups                                     1
    Number of observed variables                        12
    Number of latent factors                             4
    Estimator                                         PMLE
    Optimization method                        trust-region
    Information                                     fisher
    Strategy                                          auto
    Number of iterations (total)                      4882
    Number of two-steps (automatic)                     50
    Influence factor                                     4
    Number of parameters:
      Free                                              18
      Penalized                                         12
    Effective degrees of freedom                    24.414
    GIC                                           1896.938
    GBIC                                          1948.070
    Penalty function:
      Sparsity                                        lasso
    Optimal tuning parameter:
      Sparsity
        - Factor loadings                            0.130

  Parameter Estimates:

  Latent Variables:
                    Type  Estimate  Std.Err     2.5%      97.5%
    negative =~
      ATS_FEAR_1     pen    0.259    0.092     0.078     0.440
      ATS_FRUS_1     pen    0.004
      ATS_SAD_1      pen    0.249    0.101     0.051     0.446
      ATS_DISC_1     pen    1.066    0.193     0.687     1.445
    control =~
      ATS_INHC_1     pen    0.000
      ATS_ATTC_1     pen    0.000
    extraversion =~
      ATS_SOCI_1     pen    0.004
      ATS_HIP_1      pen    0.011
      ATS_POSA_1     pen    0.004
    sensitivity =~
      ATS_NPS_1      pen    0.307    0.113     0.086     0.528
```

ATS_APS_1	pen	0.718	0.139	0.445	0.992
ATS_ASSO_1	pen	0.451	0.118	0.219	0.683

Covariances:

	Type	Estimate	Std.Err	2.5%	97.5%
negative ~~					
control	free	-0.314	1332.806	-2612.566	2611.937
extraversion	free	-27.767	73.513	-171.851	116.316
sensitivity	free	0.382	0.143	0.101	0.663
control ~~					
extraversion	free	-0.741	122260.748	-239627.404	239625.921
sensitivity	free	-0.915	2024.741	-3969.335	3967.504
extraversion ~~					
sensitivity	free	15.537	42.629	-68.014	99.089

Variances:

	Type	Estimate	Std.Err	2.5%	97.5%
negative	fixed	1.000		1.000	1.000
control	fixed	1.000		1.000	1.000
extraversion	fixed	1.000		1.000	1.000
sensitivity	fixed	1.000		1.000	1.000
.ATS_FEAR_1	free	0.623	0.114	0.400	0.847
.ATS_FRUS_1	free	0.722	0.132	0.464	0.980
.ATS_SAD_1	free	0.957	0.173	0.617	1.297
.ATS_DISC_1	free	-0.326	0.394	-1.099	0.447
.ATS_INHC_1	free	0.607	0.111	0.390	0.825
.ATS_ATTC_1	free	0.773	0.141	0.496	1.049
.ATS_SOCI_1	free	0.972	0.177	0.625	1.320
.ATS_HIP_1	free	0.935	0.170	0.601	1.269
.ATS_POSA_1	free	0.587	0.107	0.377	0.797
.ATS_NPS_1	free	0.724	0.140	0.451	0.998
.ATS_APS_1	free	0.436	0.173	0.096	0.775
.ATS_ASSO_1	free	0.631	0.134	0.368	0.893

Next, let's apply the ridge penalty to the data. This can be done quite easily by simply replacing lasso with ridge in the pen.shrink subcommand. Everything else remains the same.

```
##RIDGE PENALTY##
penfa.ridge <- penfa(model = example.cfa.model,
data = example,
information = "fisher",
pen.shrink = "ridge",
pen.diff = "none",
eta = list(shrink = c("lambda" = 0.1),
diff = c("none" = 0)),
strategy = "auto",
verbose = FALSE)

summary(penfa.ridge)
```

The results from the ridge estimation were quite similar to those for the lasso. Notice that, as we've seen throughout the book, the degree of shrinkage associated with the ridge estimator was slightly less severe than that for the lasso. However, it's also clear that both approaches yielded qualitatively quite similar results, which would be anticipated, and which we also found with the RegSEM approach.

```
penfa 0.0.0.9000 reached convergence

    Number of observations                              60
    Number of groups                                     1
    Number of observed variables                        12
    Number of latent factors                             4
    Estimator                                         PMLE
    Optimization method                        trust-region
    Information                                     fisher
    Strategy                                          auto
    Number of iterations (total)                     11708
    Number of two-steps (automatic)                     50
    Influence factor                                     4
    Number of parameters:
      Free                                              18
      Penalized                                         12
    Effective degrees of freedom                    26.647
    GIC                                           1897.773
    GBIC                                           1953.582
    Penalty function:
      Sparsity                                         ridge
    Optimal tuning parameter:
      Sparsity
        - Factor loadings                            0.214

  Parameter Estimates:

  Latent Variables:
                    Type    Estimate  Std.Err    2.5%     97.5%
    negative =~
      ATS_FEAR_1    pen       0.312    0.106    0.105     0.520
      ATS_FRUS_1    pen       0.069    0.103   -0.132     0.270
      ATS_SAD_1     pen       0.317    0.123    0.076     0.558
      ATS_DISC_1    pen       0.886    0.142    0.608     1.164
    control =~
      ATS_INHC_1    pen       0.002
      ATS_ATTC_1    pen       0.014
    extraversion =~
      ATS_SOCI_1    pen       0.011
      ATS_HIP_1     pen       0.028
      ATS_POSA_1    pen       0.007
    sensitivity =~
      ATS_NPS_1     pen       0.369    0.124    0.126     0.612
```

ATS_APS_1	pen	0.662	0.134	0.400	0.923
ATS_ASSO_1	pen	0.499	0.124	0.257	0.742

Covariances:

	Type	Estimate	Std.Err	2.5%	97.5%
negative ~~					
control	free	3.064	61.652	-117.772	123.899
extraversion	free	-12.009	108.533	-224.731	200.712
sensitivity	free	0.452	0.160	0.138	0.766
control ~~					
extraversion	free	-490.424	10742.826	-21545.977	20565.129
sensitivity	free	-3.062	61.972	-124.525	118.401
extraversion ~~					
sensitivity	free	6.819	61.789	-114.285	127.924

Variances:

	Type	Estimate	Std.Err	2.5%	97.5%
negative	fixed	1.000		1.000	1.000
control	fixed	1.000		1.000	1.000
extraversion	fixed	1.000		1.000	1.000
sensitivity	fixed	1.000		1.000	1.000
.ATS_FEAR_1	free	0.600	0.113	0.379	0.822
.ATS_FRUS_1	free	0.715	0.131	0.459	0.971
.ATS_SAD_1	free	0.929	0.172	0.592	1.266
.ATS_DISC_1	free	0.008	0.224	-0.432	0.448
.ATS_INHC_1	free	0.607	0.111	0.390	0.824
.ATS_ATTC_1	free	0.764	0.140	0.490	1.038
.ATS_SOCI_1	free	0.971	0.177	0.624	1.319
.ATS_HIP_1	free	0.931	0.171	0.597	1.265
.ATS_POSA_1	free	0.589	0.107	0.378	0.799
.ATS_NPS_1	free	0.690	0.139	0.419	0.962
.ATS_APS_1	free	0.516	0.160	0.201	0.830
.ATS_ASSO_1	free	0.586	0.134	0.323	0.849

Structural Equation Modeling

Quite often in practice, researchers are interested in how two or more latent constructs, such as intelligence and aptitude, are related to one another. Questions of this type can quite often be addressed using latent variable models that resemble regression or path analysis. For example, a researcher might posit a model for how the theory of planned behavior relates to college students' intentions to graduate. In this case, each construct is measured by scales consisting of ordinal items on a 1 to 5 Likert-type scale. The relationships among these factors are expressed through Structural Equation Models (SEMs), which are very similar in spirit to regression models, but which involve latent rather than observed variables. Equation (7.23) expresses a general example in which factor η is the endogenous (i.e., dependent) latent variable and γ is the exogenous (i.e., independent) latent variable.

$$\eta = B\gamma + \zeta \qquad (7.23)$$

The relationship between the two latent variables is expressed in coefficient B, which is interpreted essentially in the same way as would be a regression coefficient. In addition, there exists random error, ζ, that is assumed to have a mean of 0 and variance of ϕ.

The relationships expressed in (7.23) are typically referred to as the structural part of the model, whereas the linking of observed indicators to the latent variables (η and γ) is referred to as the measurement portion of the model. The statistical mechanisms for fitting these models and obtaining parameter estimates, as well as the tools used to assess model fit are much the same as for fitting and assessing CFA solutions. Therefore, a great deal of what we see in this chapter will be similar to what we discussed earlier.

The fitting of SEMs occurs in two steps. First, the CFA models for the latent factors must be estimated and shown to fit the data well. In doing this, we are ensuring that the latent variables that we propose for inclusion in the structural model are in fact supported by the data. If the measurement model does not fit well, then it is not possible to go forward and estimate the structural component because the factors themselves are not dependable. Assuming that the measurement models do in fact fit the data, we can then proceed to fit the structural components. Here, our interest is in determining whether the hypothesized model(s) expressing relationships among the factors fit the data, and in identifying the nature of the relationships themselves. As noted earlier, determination of model fit is done in much the same way as it is for CFA, using various indices such as RMSEA, CFI, TLI, SRMR, and the chi-square goodness of fit test. In addition, we will obtain estimates for the coefficients linking the latent variables along with standard errors and hypothesis tests.

We'll begin our examination of fitting SEM in R by using a diagonally weighted least squares estimator to fit a model in which the latent variable coping serves as the response, and the predictors are negative, control, extraversion, and sensitivity. We can fit this model using lavaan, as below. Note that the structure of these commands is quite similar to that of the CFA model. The additional piece for SEM is the inclusion of the structural model in which coping serves as the dependent latent variable, and negative affect, effortful control, extraversion/surgency, and orienting sensitivity are the independent variables.

```
# fit standard SEM#
example.sem.model <- ' negative =~ NA*ATS_FEAR_1 + ATS_FRUS_1 +
    ATS_SAD_1 + ATS_DISC_1
    control =~  NA*ATS_INHC_1  + ATS_ATTC_1
    extraversion =~ NA*ATS_SOCI_1 + ATS_HIP_1 + ATS_POSA_1
    sensitivity =~ NA*ATS_NPS_1 + ATS_APS_1 + ATS_ASSO_1
    coping =~ NA*SCQ_DEN_1+SCQ_HUM_1 +
```

```
         SCQ_ACT_1+SCQ_PEER_1+SCQ_CON_1+SCQ_HELP_1+SCQ_POP_1
         negative~~1*negative
         control~~1*control
         extraversion~~1*extraversion
         sensitivity~~1*sensitivity
         coping~~1*coping
         coping ~ negative + control + extraversion + sensitivity'
```

example.sem.results <- sem(example.sem.model, data=example, estimator="DWLS")

summary(example.sem.results, fit.measures=T, standardized=T)

lavaan 0.6-6 ended normally after 134 iterations

```
  Estimator                                         DWLS
  Optimization method                             NLMINB
  Number of free parameters                           48
  Number of observations                              60
Model Test User Model:
  Test statistic                                 150.108
  Degrees of freedom                                 142
  P-value (Chi-square)                             0.304

Model Test Baseline Model:

  Test statistic                                 374.515
  Degrees of freedom                                 171
  P-value                                          0.000

User Model versus Baseline Model:

  Comparative Fit Index (CFI)                      0.960
  Tucker-Lewis Index (TLI)                         0.952

Root Mean Square Error of Approximation:

  RMSEA                                            0.031
  90 Percent confidence interval - lower           0.000
  90 Percent confidence interval - upper           0.070
  P-value RMSEA <= 0.05                            0.742

Standardized Root Mean Square Residual:

  SRMR                                             0.113

Parameter Estimates:

  Standard errors                               Standard
  Information                                   Expected
  Information saturated (h1) model          Unstructured

Latent Variables:
               Estimate  Std.Err  z-value  P(>|z|)  Std.lv  Std.all
  negative =~
    ATS_FEAR_1    0.502    0.093    5.425    0.000   0.502    0.575
    ATS_FRUS_1    0.261    0.091    2.853    0.004   0.261    0.304
```

ATS_SAD_1	0.794	0.135	5.872	0.000	0.794	0.752
ATS_DISC_1	0.628	0.111	5.642	0.000	0.628	0.621
control =~						
ATS_INHC_1	0.172	0.161	1.073	0.283	0.172	0.219
ATS_ATTC_1	0.260	0.242	1.073	0.283	0.260	0.293
extraversion =~						
ATS_SOCI_1	0.439	0.147	2.987	0.003	0.439	0.441
ATS_HIP_1	0.298	0.114	2.607	0.009	0.298	0.301
ATS_POSA_1	0.281	0.096	2.933	0.003	0.281	0.362
sensitivity =~						
ATS_NPS_1	0.700	0.115	6.084	0.000	0.700	0.732
ATS_APS_1	0.690	0.121	5.718	0.000	0.690	0.638
ATS_ASSO_1	0.551	0.106	5.213	0.000	0.551	0.560
coping =~						
SCQ_DEN_1	0.015	0.805	0.019	0.985	0.117	0.122
SCQ_HUM_1	-0.060	3.115	-0.019	0.985	-0.452	-0.431
SCQ_ACT_1	-0.077	4.015	-0.019	0.985	-0.583	-0.588
SCQ_PEER_1	0.027	1.423	0.019	0.985	0.207	0.227
SCQ_CON_1	0.007	0.386	0.019	0.985	0.056	0.053
SCQ_HELP_1	-0.083	4.304	-0.019	0.985	-0.625	-0.639
SCQ_POP_1	-0.018	0.958	-0.019	0.985	-0.139	-0.124

Regressions:

	Estimate	Std.Err	z-value	P(>|z|)	Std.lv	Std.all
coping ~						
negative	3.614	171.512	0.021	0.983	0.479	0.479
control	7.124	341.398	0.021	0.983	0.945	0.945
extraversion	1.049	24.927	0.042	0.966	0.139	0.139
sensitivity	-5.277	260.219	-0.020	0.984	-0.700	-0.700

Covariances:

	Estimate	Std.Err	z-value	P(>|z|)	Std.lv	Std.all
negative ~~						
control	-0.333	0.426	-0.782	0.434	-0.333	-0.333
extraversion	-0.061	0.166	-0.365	0.715	-0.061	-0.061
sensitivity	0.503	0.113	4.448	0.000	0.503	0.503
control ~~						
extraversion	-1.080	1.138	-0.950	0.342	-1.080	-1.080
sensitivity	-0.274	0.421	-0.650	0.516	-0.274	-0.274
extraversion ~~						
sensitivity	0.407	0.218	1.868	0.062	0.407	0.407

Variances:

	Estimate	Std.Err	z-value	P(>|z|)	Std.lv	Std.all
negative	1.000				1.000	1.000
control	1.000				1.000	1.000
extraversion	1.000				1.000	1.000
sensitivity	1.000				1.000	1.000
.coping	1.000				0.018	0.018
.ATS_FEAR_1	0.510	0.159	3.213	0.001	0.510	0.669
.ATS_FRUS_1	0.667	0.126	5.309	0.000	0.667	0.907

.ATS_SAD_1	0.485	0.276	1.754	0.079	0.485	0.435
.ATS_DISC_1	0.630	0.202	3.113	0.002	0.630	0.615
.ATS_INHC_1	0.588	0.125	4.707	0.000	0.588	0.952
.ATS_ATTC_1	0.718	0.174	4.123	0.000	0.718	0.914
.ATS_SOCI_1	0.800	0.203	3.938	0.000	0.800	0.806
.ATS_HIP_1	0.891	0.171	5.194	0.000	0.891	0.910
.ATS_POSA_1	0.522	0.123	4.248	0.000	0.522	0.869
.ATS_NPS_1	0.423	0.227	1.866	0.062	0.423	0.464
.ATS_APS_1	0.693	0.259	2.673	0.008	0.693	0.593
.ATS_ASSO_1	0.666	0.179	3.723	0.000	0.666	0.687
.SCQ_DEN_1	0.901	0.149	6.047	0.000	0.901	0.985
.SCQ_HUM_1	0.896	0.210	4.258	0.000	0.896	0.814
.SCQ_ACT_1	0.644	0.182	3.549	0.000	0.644	0.655
.SCQ_PEER_1	0.784	0.140	5.590	0.000	0.784	0.948
.SCQ_CON_1	1.096	0.175	6.250	0.000	1.096	0.997
.SCQ_HELP_1	0.565	0.200	2.820	0.005	0.565	0.592
.SCQ_POP_1	1.240	0.170	7.299	0.000	1.240	0.985

We will focus on the structural coefficients linking the factors to one another. However, it is worth first examining the factor loadings in order to assess the utility of the latent variables themselves. Perhaps most notable is that the indicators believed to be associated with the coping latent variable are quite small, which is likely to impact the estimation of the structural coefficients. An examination of these coefficients reveals that there does not appear to be a relationship between any of the predictor factors and coping, given the non-significant hypothesis test results for each.

RegSEM Structural Model

We can fit the SEM above using the lasso estimator using the regsem library. First, we will apply cross-validation to identify the optimal penalty parameter. As with the CFA models above, we will investigate 100 potential λ, with jumps of 0.05 between adjacent values. We will penalize both factor loadings and the structure coefficients (regressions) linking the response and predictor latent variables. The specification search is based on results for the optimal penalty value on the results of the maximum likelihood lavaan results (example.sem. results.ml). We will then plot the parameter estimates by the regularization penalty value, showing the optimal value based on the value of BIC.

```
#FIT LASSO SEM#
lasso.cv.results <- cv_regsem(example.sem.results.
ml,type="lasso", pars_pen=c("loadings","regressions"),
gradFun="ram", n.lambda=100, jump=0.05)

plot(lasso.cv.results, show.minimum="BIC")
```

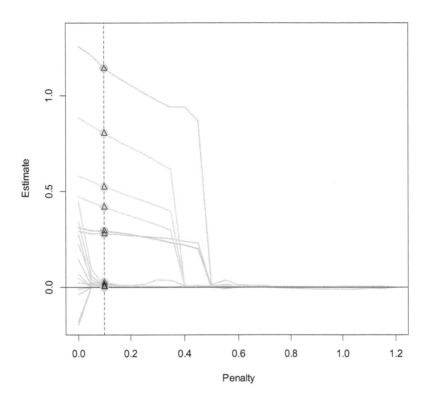

FIGURE 7.6
Factor loading and structure coefficient estimates by penalty parameter value

From this plot, we can see that the algorithm searched a sufficient number of penalty values that all coefficients were shrunk to 0. The optimal λ based on the BIC was approximately 0.1. We can obtain the optimal value by using the summary function.

```
summary(lasso.cv.results)
CV regsem Object
  Number of parameters regularized: 19
  Lambda ranging from 0 to 1.2
  Lowest Fit Lambda: 0.1
  Metric: BIC
  Number Converged: 25
```

For this example, the optimal regularization tuning parameter was indeed 0.1. The model parameters associated with this optimal λ value appear below from the lasso.cv.results$final _ pars command. As for the CFA results that were described earlier, a number of the factor loadings are either 0 or very close to it. The coefficients linking the individual factors to coping are extremely large, which is likely due to the fact that the factor loadings for coping were all regularized.

```
lasso.cv.results$final_pars
```

negative -> ATS_FEAR_1	negative -> ATS_FRUS_1	negative -> ATS_SAD_1
0.278	0.014	0.293
negative -> ATS_DISC_1	control -> ATS_INHC_1	control -> ATS_ATTC_1
1.145	0.006	0.024
extraversion -> ATS_SOCI_1	extraversion -> ATS_HIP_1	extraversion -> ATS_POSA_1
0.001	0.001	0.000
sensitivity -> ATS_NPS_1	sensitivity -> ATS_APS_1	sensitivity -> ATS_ASSO_1
0.417	0.803	0.523
coping -> SCQ_DEN_1	coping -> SCQ_HUM_1	coping -> SCQ_ACT_1
0.000	0.000	0.000
coping -> SCQ_PEER_1	coping -> SCQ_CON_1	coping -> SCQ_HELP_1
0.000	0.000	0.000
coping -> SCQ_POP_1	negative -> coping	control -> coping
0.000	-1982.467	2557.034
extraversion -> coping	sensitivity -> coping	ATS_FEAR_1 ~~ ATS_FEAR_1
-31.369	-1873.953	0.645
ATS_FRUS_1 ~~ ATS_FRUS_1	ATS_SAD_1 ~~ ATS_SAD_1	ATS_DISC_1 ~~ ATS_DISC_1
0.721	0.971	-0.420
ATS_INHC_1 ~~ ATS_INHC_1	ATS_ATTC_1 ~~ ATS_ATTC_1	ATS_SOCI_1 ~~ ATS_SOCI_1
0.607	0.765	0.970
ATS_HIP_1 ~~ ATS_HIP_1	ATS_POSA_1 ~~ ATS_POSA_1	ATS_NPS_1 ~~ ATS_NPS_1
0.947	0.587	0.676
ATS_APS_1 ~~ ATS_APS_1	ATS_ASSO_1 ~~ ATS_ASSO_1	SCQ_DEN_1 ~~ SCQ_DEN_1
0.402	0.618	0.875
SCQ_HUM_1 ~~ SCQ_HUM_1	SCQ_ACT_1 ~~ SCQ_ACT_1	SCQ_PEER_1 ~~ SCQ_PEER_1
0.805	0.652	0.768
SCQ_CON_1 ~~ SCQ_CON_1	SCQ_HELP_1 ~~ SCQ_HELP_1	SCQ_POP_1 ~~ SCQ_POP_1
1.080	0.571	1.221
negative ~~ control	negative ~~ extraversion	negative ~~ sensitivity
-1.135	-230.958	0.370
control ~~ extraversion	control ~~ sensitivity	extraversion ~~ sensitivity
-4450.056	-3.420	144.821

Now let's apply the lasso penalty only to the regression coefficients, and let the loadings be freely estimated. This is done using the pars _ pen subcommand with only regressions specified, as follows:

```
lasso.cv.results <- cv_regsem(example.sem.results.
ml,type="lasso", pars_pen=c("regressions"), gradFun="ram",
n.lambda=100, jump=0.05)

plot(lasso.cv.results, show.minimum="BIC")
```

The optimal value of the regularization tuning parameter appears to be 0.35, based on the both graph (above) and based on the summary command (below).

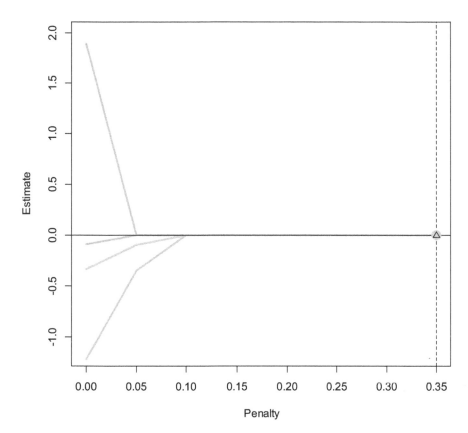

FIGURE 7.7
Structure coefficient estimates by penalty parameter value

```
summary(lasso.cv.results)
CV regsem Object
 Number of parameters regularized: 4
 Lambda ranging from 0 to 0.35
 Lowest Fit Lambda: 0.35
 Metric: BIC
 Number Converged: 8
```

The parameter estimates associated with the optimal λ value appear below. First, note that the factor loadings for the coping indicators are no longer regularized to 0. Second, the coefficients linking the latent predictors to coping were all regularized to 0. These more stable structural coefficient estimates are likely due to the fact that the factor loadings for coping have not been shrunken to 0, as in the previous example.

```
lasso.cv.results$final_pars
```

negative -> ATS_FEAR_1	negative -> ATS_FRUS_1	negative -> ATS_SAD_1
0.294	0.024	0.314
negative -> ATS_DISC_1	control -> ATS_INHC_1	control -> ATS_ATTC_1
1.255	0.000	0.000
extraversion -> ATS_SOCI_1	extraversion -> ATS_HIP_1	extraversion -> ATS_POSA_1
0.344	0.444	0.227
sensitivity -> ATS_NPS_1	sensitivity -> ATS_APS_1	sensitivity -> ATS_ASSO_1
0.472	0.891	0.579
coping -> SCQ_DEN_1	coping -> SCQ_HUM_1	coping -> SCQ_ACT_1
0.164	-0.488	-0.601
coping -> SCQ_PEER_1	coping -> SCQ_CON_1	coping -> SCQ_HELP_1
0.176	-0.066	-0.649
coping -> SCQ_POP_1	negative -> coping	control -> coping
-0.125	0.000	0.000
extraversion -> coping	sensitivity -> coping	ATS_FEAR_1 ~~ ATS_FEAR_1
0.000	0.000	0.663
ATS_FRUS_1 ~~ ATS_FRUS_1	ATS_SAD_1 ~~ ATS_SAD_1	ATS_DISC_1 ~~ ATS_DISC_1
0.722	0.998	-0.565
ATS_INHC_1 ~~ ATS_INHC_1	ATS_ATTC_1 ~~ ATS_ATTC_1	ATS_SOCI_1 ~~ ATS_SOCI_1
0.607	0.773	0.852
ATS_HIP_1 ~~ ATS_HIP_1	ATS_POSA_1 ~~ ATS_POSA_1	ATS_NPS_1 ~~ ATS_NPS_1
0.755	0.537	0.675
ATS_APS_1 ~~ ATS_APS_1	ATS_ASSO_1 ~~ ATS_ASSO_1	SCQ_DEN_1 ~~ SCQ_DEN_1
0.358	0.620	0.873
SCQ_HUM_1 ~~ SCQ_HUM_1	SCQ_ACT_1 ~~ SCQ_ACT_1	SCQ_PEER_1 ~~ SCQ_PEER_1
0.845	0.608	0.783
SCQ_CON_1 ~~ SCQ_CON_1	SCQ_HELP_1 ~~ SCQ_HELP_1	SCQ_POP_1 ~~ SCQ_POP_1
1.077	0.521	1.223
negative ~~ control	negative ~~ extraversion	negative ~~ sensitivity
274.768	-0.632	0.350
control ~~ extraversion	control ~~ sensitivity	extraversion ~~ sensitivity
-6625.584	-3873.474	0.331

The regsem library allows for applying the ridge estimator to SEM just as we can for CFA. The command structure is identical for our lasso example, except that we specify type="ridge".

```
#FIT RIDGE SEM#
ridge.cv.results <- cv_regsem(example.sem.results.
ml,type="ridge", pars_pen=c("loadings","regressions"),
gradFun="ram", n.lambda=100, jump=0.05)

plot(ridge.cv.results, show.minimum="BIC")
```

Based on the plot, it appears that the optimal penalty value is approximately 0.1. The summary function reveals that in fact the optimal λ is 0.1. The model parameters associated with this tuning parameter value appear below, and are very similar to those from the lasso estimator.

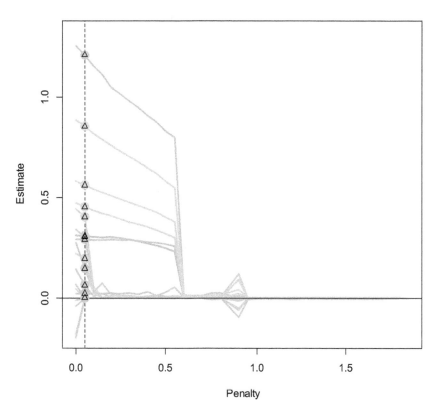

FIGURE 7.8

Factor loading and structure coefficient estimates by penalty parameter value

```
ridge.cv.results$final_pars
```

negative -> ATS_FEAR_1	negative -> ATS_FRUS_1	negative -> ATS_SAD_1
0.291	0.022	0.309
negative -> ATS_DISC_1	control -> ATS_INHC_1	control -> ATS_ATTC_1
1.213	0.064	0.147
extraversion -> ATS_SOCI_1	extraversion -> ATS_HIP_1	extraversion -> ATS_POSA_1
0.304	0.404	0.197
sensitivity -> ATS_NPS_1	sensitivity -> ATS_APS_1	sensitivity -> ATS_ASSO_1
0.454	0.855	0.563
coping -> SCQ_DEN_1	coping -> SCQ_HUM_1	coping -> SCQ_ACT_1
0.000	0.000	0.000
coping -> SCQ_PEER_1	coping -> SCQ_CON_1	coping -> SCQ_HELP_1
0.000	0.000	0.000
coping -> SCQ_POP_1	negative -> coping	control -> coping
0.000	-1438.857	1345.217
extraversion -> coping	sensitivity -> coping	ATS_FEAR_1 ~~ ATS_FEAR_1
-2035.668	-808.412	0.655
ATS_FRUS_1 ~~ ATS_FRUS_1	ATS_SAD_1 ~~ ATS_SAD_1	ATS_DISC_1 ~~ ATS_DISC_1
0.722	0.988	-0.506

ATS_INHC_1 ~~ ATS_INHC_1	ATS_ATTC_1 ~~ ATS_ATTC_1	ATS_SOCI_1 ~~ ATS_SOCI_1
0.602	0.746	0.872
ATS_HIP_1 ~~ ATS_HIP_1	ATS_POSA_1 ~~ ATS_POSA_1	ATS_NPS_1 ~~ ATS_NPS_1
0.782	0.547	0.674
ATS_APS_1 ~~ ATS_APS_1	ATS_ASSO_1 ~~ ATS_ASSO_1	SCQ_DEN_1 ~~ SCQ_DEN_1
0.378	0.614	0.879
SCQ_HUM_1 ~~ SCQ_HUM_1	SCQ_ACT_1 ~~ SCQ_ACT_1	SCQ_PEER_1 ~~ SCQ_PEER_1
0.780	0.659	0.773
SCQ_CON_1 ~~ SCQ_CON_1	SCQ_HELP_1 ~~ SCQ_HELP_1	SCQ_POP_1 ~~ SCQ_POP_1
1.079	0.562	1.222
negative ~~ control	negative ~~ extraversion	negative ~~ sensitivity
0.405	-0.684	0.351
control ~~ extraversion	control ~~ sensitivity	extraversion ~~ sensitivity
-2.572	-0.416	0.373

Now, let's fit the model and only regularize the structural coefficients using the ridge estimator.

```
ridge.cv.results <- cv_regsem(example.sem.results.
ml,type="ridge", pars_pen=c("regressions"), gradFun="ram",
n.lambda=100, jump=0.05)

plot(ridge.cv.results, show.minimum="BIC")
```

The optimal tuning parameter was 0.35, which is the same as for the lasso estimator. The model parameters appear below, and are very close to those from the lasso.

```
ridge.cv.results$final_pars
```

negative -> ATS_FEAR_1	negative -> ATS_FRUS_1	negative -> ATS_SAD_1
0.290	0.022	0.311
negative -> ATS_DISC_1	control -> ATS_INHC_1	control -> ATS_ATTC_1
1.266	0.000	0.000
extraversion -> ATS_SOCI_1	extraversion -> ATS_HIP_1	extraversion -> ATS_POSA_1
0.340	0.438	0.225
sensitivity -> ATS_NPS_1	sensitivity -> ATS_APS_1	sensitivity -> ATS_ASSO_1
0.468	0.892	0.574
coping -> SCQ_DEN_1	coping -> SCQ_HUM_1	coping -> SCQ_ACT_1
0.168	-0.487	-0.598
coping -> SCQ_PEER_1	coping -> SCQ_CON_1	coping -> SCQ_HELP_1
0.182	-0.061	-0.647
coping -> SCQ_POP_1	negative -> coping	control -> coping
-0.127	0.000	0.001
extraversion -> coping	sensitivity -> coping	ATS_FEAR_1 ~~ ATS_FEAR_1
0.000	0.000	0.665
ATS_FRUS_1 ~~ ATS_FRUS_1	ATS_SAD_1 ~~ ATS_SAD_1	ATS_DISC_1 ~~ ATS_DISC_1
0.722	1.000	-0.594
ATS_INHC_1 ~~ ATS_INHC_1	ATS_ATTC_1 ~~ ATS_ATTC_1	ATS_SOCI_1 ~~ ATS_SOCI_1
0.607	0.773	0.856

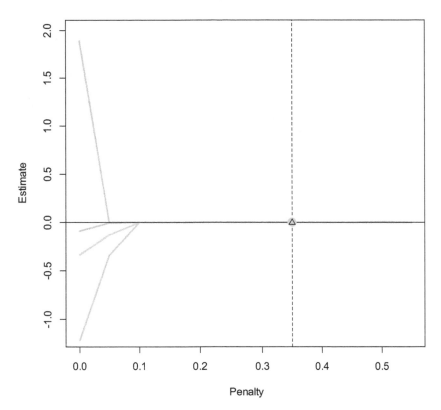

FIGURE 7.9
Structure coefficient estimates by penalty parameter value

ATS_HIP_1 ~~ ATS_HIP_1	ATS_POSA_1 ~~ ATS_POSA_1	ATS_NPS_1 ~~ ATS_NPS_1
0.763	0.538	0.678
ATS_APS_1 ~~ ATS_APS_1	ATS_ASSO_1 ~~ ATS_ASSO_1	SCQ_DEN_1 ~~ SCQ_DEN_1
0.349	0.622	0.871
SCQ_HUM_1 ~~ SCQ_HUM_1	SCQ_ACT_1 ~~ SCQ_ACT_1	SCQ_PEER_1 ~~ SCQ_PEER_1
0.846	0.611	0.780
SCQ_CON_1 ~~ SCQ_CON_1	SCQ_HELP_1 ~~ SCQ_HELP_1	SCQ_POP_1 ~~ SCQ_POP_1
1.078	0.523	1.222
negative ~~ control	negative ~~ extraversion	negative ~~ sensitivity
72.654	-0.637	0.343
control ~~ extraversion	control ~~ sensitivity	extraversion ~~ sensitivity
-1203.415	-650.243	0.327

2-Stage Least Squares Estimation for SEM

An alternative method for parameter estimation that has been found to be particularly useful for addressing the problem of model misspecification is

2-stage least squares (2SLS). Bollen (1996) proposed use of this method in the context of SEM, and subsequent work has demonstrated its utility in this context. The 2SLS estimation approach reexpresses the latent variable model in terms of a model that directly relates the observed indicator variables with one another. In the context of SEM, each indicator can be expressed in terms of the latent variable to which it is related, and random error, as exhibited later. Note that these equations assume univariate η and ξ, although the 2SLS estimator can easily accommodate more complex models with no loss of generality.

$$y_1 = \eta + \epsilon_1 \tag{7.24}$$
$$x_1 = \xi + \delta_1 \tag{7.25}$$

where
η = Endogenous latent variable(s)
ξ = Exogenous latent variable(s)
y_1 = Referent indicator variable(s) for η
x_1 = Referent indicator variable(s) for ξ
ϵ_1 = Error term for y_1
δ_1 = Error term for x_1

Through simple substitution, we can obtain the following:

$$\eta = y_1 - \epsilon_1 \tag{7.26}$$
$$\xi = x_1 - \delta_1 \tag{7.27}$$

Now that the latent variables are expressed in terms of the observed indicators, the SEM (assuming only one endogenous variable) in equation (1) can be rewritten in terms of the observed indicators.

$$y_1 = \Gamma x_1 + u \tag{7.28}$$

where
$u = \epsilon_1 - \Gamma \delta_1 + \zeta$; i.e., the sum of the disturbance terms.

This model involves only observed variables in place of the latent factors, and thus theoretically is estimable using observed variable methods. However, the model is problematic in practice because x_1 is correlated with terms in u. Therefore, a commonly used approach for fitting linear models, ordinary least squares (OLS), is not appropriate given its assumption of the error term and independent variables being uncorrelated is violated. A popular strategy for addressing this estimation issue is to use an approach based upon instrumental variables (IV). An IV is simply a variable that is correlated with x_1, in this case, but is uncorrelated with u. Therefore, in the first stage of 2SLS one or more IVs are selected and then used as independent variables with a regression model in which x_1 serves as the dependent variable. For each member of the sample, a predicted value of x_1, $\hat{x}_1$ is then calculated, based upon the regression model using the IV(s). This new variable is no longer correlated with the error term but does retain whatever relationship it had with y_1. In this way, the problem of dependence inherent in equation (11) has been removed. Bollen and Bauer (2004) demonstrated that in the context of SEM, IVs can be selected from among the non-referent indicators for a factor. In the second stage of 2SLS model equation (12) can be rewritten as

$$y_1 = \Gamma\hat{x}_1 + u \qquad (7.29)$$

Finally, standard errors of the parameter estimates are obtained using the bootstrap. Determinations regarding the statistical significance of the predictors can be made using confidence intervals based upon this bootstrap approach.

Standard 2SLS with R

First, let's fit the standard 2SLS estimator to our ATS data using the MIIVsem library in R. We must first define the model, which takes a similar form as with lavaan. We then use the miive function to obtain parameter estimates using 2SLS. The summary function then provides us with the model parameter estimates.

```
library(MIIVsem)

example.2sls.model <- ' negative =~ ATS_FEAR_1 + ATS_FRUS_1
+ ATS_SAD_1 + ATS_DISC_1
    control =~  ATS_INHC_1  + ATS_ATTC_1
    extraversion =~ ATS_SOCI_1 + ATS_HIP_1 + ATS_POSA_1
    sensitivity =~ ATS_NPS_1 + ATS_APS_1 + ATS_ASSO_1
    coping =~ SCQ_DEN_1+SCQ_HUM_1 + SCQ_ACT_1+SCQ_
PEER_1+SCQ_CON_1+SCQ_HELP_1+SCQ_POP_1
    coping ~ negative + control + extraversion + sensitivity'

miiv.results<-miive(model = example.2sls.model, data = example)

summary(miiv.results)

MIIVsem (0.5.5) results

Number of observations                                     60
Number of equations                                       15
Estimator                                          MIIV-2SLS
Standard Errors                                     standard
Missing                                             listwise
```

```
Parameter Estimates:

STRUCTURAL COEFFICIENTS:
```

	Estimate	Std.Err	z-value	P(>|z|)	Sargan	df	P(Chi)
control =~							
ATS_INHC_1	1.000						
ATS_ATTC_1	-0.003	0.300	-0.010	0.992	16.038	16	0.450
coping =~							
SCQ_DEN_1	1.000						
SCQ_HUM_1	-0.243	0.282	-0.860	0.390	30.223	16	0.017
SCQ_ACT_1	-0.291	0.260	-1.117	0.264	28.745	16	0.026
SCQ_PEER_1	0.110	0.238	0.461	0.645	28.232	16	0.030
SCQ_CON_1	0.060	0.273	0.218	0.828	35.331	16	0.004
SCQ_HELP_1	-0.231	0.265	-0.870	0.384	27.771	16	0.034
SCQ_POP_1	-0.434	0.311	-1.397	0.162	15.277	16	0.504
extraversion =~							
ATS_SOCI_1	1.000						
ATS_HIP_1	0.499	0.229	2.175	0.030	23.424	16	0.103
ATS_POSA_1	0.646	0.207	3.130	0.002	4.560	16	0.998
negative =~							
ATS_FEAR_1	1.000						
ATS_FRUS_1	0.534	0.199	2.687	0.007	22.251	16	0.135
ATS_SAD_1	0.923	0.251	3.676	0.000	18.602	16	0.290
ATS_DISC_1	0.943	0.257	3.664	0.000	26.187	16	0.051
sensitivity =~							
ATS_NPS_1	1.000						
ATS_APS_1	0.610	0.207	2.951	0.003	30.505	16	0.016
ATS_ASSO_1	0.514	0.186	2.757	0.006	26.198	16	0.051
coping ~							
negative	0.568	0.306	1.858	0.063	1.963	4	0.743
control	0.700	0.613	1.142	0.253			
extraversion	0.074	0.437	0.170	0.865			
sensitivity	-0.052	0.336	-0.156	0.876			

```
INTERCEPTS:
```

	Estimate	Std.Err	z-value	P(>|z|)
ATS_APS_1	1.856	0.980	1.894	0.058
ATS_ASSO_1	2.605	0.884	2.948	0.003
ATS_ATTC_1	4.107	1.314	3.125	0.002
ATS_DISC_1	0.472	1.000	0.472	0.637
ATS_FEAR_1	0.000			
ATS_FRUS_1	1.939	0.773	2.507	0.012
ATS_HIP_1	1.806	1.085	1.665	0.096
ATS_INHC_1	0.000			
ATS_NPS_1	0.000			
ATS_POSA_1	1.519	0.977	1.555	0.120
ATS_SAD_1	0.706	0.978	0.721	0.471
ATS_SOCI_1	0.000			
coping	-1.205	3.877	-0.311	0.756
SCQ_ACT_1	5.803	1.086	5.342	0.000

SCQ_CON_1	3.323	1.141	2.911	0.004
SCQ_DEN_1	0.000			
SCQ_HELP_1	6.401	1.106	5.787	0.000
SCQ_HUM_1	4.895	1.177	4.158	0.000
SCQ_PEER_1	3.065	0.995	3.081	0.002
SCQ_POP_1	6.295	1.297	4.853	0.000

As with the ML estimator, our focus will be on the structural coefficients in this case, though it is also good practice to examine the factor loadings as well. If the loadings do not reflect the presence of strong latent variables (as is the case here), then we will want to interpret the structural coefficients with caution. One aspect of 2SLS that is not an issue with standard SEM is the use of instrumental variables. Sargan's statistic tests the null hypothesis that the overidentification restriction associated with the instruments for each variable is met. Thus, a statistically significant result would indicate that, for that variable, the instruments are not appropriate. We see that the structural portion of the model does meet this criterion. In terms of the structural relationships, it appears that none of the independent latent traits are associated with coping.

Regularized 2SLS

The 2SLS approach to modeling SEMs has been extended to incorporate regularization techniques (Jung, 2013). Prior work has demonstrated that 2SLS estimates can be biased when samples are small, particularly when the number of IVs is large (Johnston, 1984). Jung's solution to this problem was to use regularization to yield more stable estimates in the presence of small samples, while taking advantage of the positive aspects of 2SLS as outlined earlier. This approach is based upon the well-known regularization approaches used in ordinary least squares regression. As an example, the 2SLS Ridge function can be written as:

$$f_{ridge} = \sum_{i=1}^{n}(w_i - Z_i a_i)^2 + \lambda \sum_{j=1}^{p} a_i^2 \qquad (7.30)$$

where

w_i = Vector of observations on the dependent variable y_1
Z_i = Vector of predictor variables from y_1 and x_1
a_i = Model coefficient estimates
λ = Tuning parameter

The 2SLS Ridge estimator is essentially just the Ridge estimator that is used in standard regression problems (e.g., Hoerl & Kennard, 1970). Given that

2SLS relies on the regression framework, other regularization methods applicable in that context, such as the lasso and elastic net techniques, can also be used much as 2SLS Ridge could. The 2SLS Lasso is based upon the Lasso estimator described by Tibshirani (1996) and can be expressed as:

$$f_{lasso} = \sum_{i=1}^{n}(w_i - Z_i a_i)^2 + \lambda\sum_{j=1}^{p}|a_i| \qquad (7.31)$$

Likewise, the 2SLS Elastic Net (2SLS Net) function as defined by Jung is:

$$f_{enet} = \sum_{i=1}^{n}(w_i - Z_i a_i)^2 + \lambda\left((1-v)\sum_{j=1}^{p}a_i^2 + v\sum_{j=1}^{p}|a_i|\right) \qquad (7.32)$$

where
   v = Mixing parameter between ridge ($v = 0$) and lasso ($v = 1$).

Both the RegSEM approach described earlier, and the regularized 2SLS described here rely on the same basic framework for estimating model parameters. Namely, they are designed to minimize the difference between model estimated values and the actual data, with a penalty applied to model parameter values. However, these two regularization frameworks also differ with respect to the criteria that are used to determine the parameter estimates. For example, as RegSEM is based on ML estimation, its focus is on minimizing the difference between the model implied and observed covariance matrices among the variables included in the model while including a penalty for model parameter estimates (equation 4). The regularized 2SLS estimator, on the other hand, is designed to minimize the difference between the actual value of the dependent indicator variable and the model predicted value, with a regularization penalty applied (equations 14, 15, and 16). Thus, the focus in RegSEM is on the entire set of model parameters, whereas for regularized 2SLS it is on the relationships between the independent and dependent latent variables, as they are represented by their observed indicators.

   As with the RegSEM approaches described earlier, appropriate values for λ and α must be selected for each of the three forms of the regularized 2SLS. K- fold cross-validation is one of the more popular methods for determining the α and λ for each model (cf. Friedman, Hastie, & Tibshirani, 2010). In cross-validation, the sample is divided into K different folds from 5,10, or n. Next, a value for the tuning parameter λ is chosen, and the model is then fit on the $K=1$ fold with the regularization estimated coefficients. These coefficient values are used to predict values in the remaining k folds. The process is then repeated for each of the remaining $k = 2, \ldots . K$ folds. A measure of fit (e.g., root mean squared error) is recorded for each value of k, and averages and standard errors for the measure of it are calculated for all folds for a particular value of λ. This process is iterated a large number of times (e.g., 100) for different values of λ. The model with the best fit measurement (i.e., lowest root mean squared error) is determined, and its corresponding tuning parameter is chosen. A similar process is repeated for elastic net for both λ and α.

Fitting the 2SLS lasso estimator to the data involves multiple steps. First, we must create matrices containing the instrumental variable for each indicator, and then fit penalized models for each of the latent traits, in turn, using one indicator as the dependent variable. Next, we use cross-validation to identify the optimal value of the tuning parameter for each latent trait. These optimal values are then saved and used to obtain predicted values of the indicators based on their instruments. Finally, we take the predicted values for our independent indicator variables and use them in a regression model for the coping dependent variable, and use the summary command to obtain the structure coefficients.

```
#Lasso 2sls#

#CREATE THE INSTRUMENTAL VARIABLE SETS FOR EACH INDICATOR#
example.x.negative<-data.frame(example$ATS_FRUS_1,
example$ATS_SAD_1,example$ATS_DISC_1)
example.x.extraversion<-data.frame(example$ATS_SOCI_1,
example$ATS_HIP_1,example$ATS_POSA_1)
example.x.control<-data.frame(example$ATS_ACTC_1,
example$ATS_INHC_1)
example.x.sensitivity<-data.frame(example$ATS_APS_1,
example$ATS_ASSO_1)

#Stage 1#
lasso.2sls.negative<-glmnet(as.matrix(example.x.negative),e
xample$ATS_FEAR_1,alpha=1, nlambda=100)
lasso.2sls.extraversion<-glmnet(as.matrix(example.x.extrave
rsion),example$ATS_SOCI_1,alpha=1, nlambda=100)
lasso.2sls.control<-glmnet(as.matrix(example.x.control),exa
mple$ATS_ACTC_1,alpha=1, nlambda=100)
lasso.2sls.sensitivity<-glmnet(as.matrix(example.x.sensitiv
ity),example$ATS_NPS_1,alpha=1, nlambda=100)

#Identify optimal tuning parameter values#
lasso.2sls.negative.cv<-cv.glmnet(as.matrix(example.x.negat
ive),example$ATS_FEAR_1)
lasso.2sls.extraversion.cv<-cv.glmnet(as.matrix(example.x.e
xtraversion),example$ATS_SOCI_1)
lasso.2sls.control.cv<-cv.glmnet(as.matrix(example.x.contro
l),example$ATS_ACTC_1)
lasso.2sls.sensitivity.cv<-cv.glmnet(as.matrix(example.x.se
nsitivity),example$ATS_NPS_1)

#Save optimal tuning parameter values#
lasso.2sls.negative.lambda<-lasso.2sls.negative.cv$lambda.min
lasso.2sls.extraversion.lambda<-lasso.2sls.extraversion.
```

```
cv$lambda.min
lasso.2sls.control.lambda<-lasso.2sls.control.cv$lambda.min
lasso.2sls.sensitivity.lambda<-lasso.2sls.sensitivity.
cv$lambda.min

#Obtain predicted values based on instruments#
lasso.2sls.negative.predict<-predict(lasso.2sls.negative,
as.matrix(example.x.negative), s=lasso.2sls.negative.lambda)
lasso.2sls.extraversion.predict<-predict(lasso.2sls.
extraversion, as.matrix(example.x.extraversion),
s=lasso.2sls.extraversion.lambda)
lasso.2sls.control.predict<-predict(lasso.2sls.control,
as.matrix(example.x.control), s=lasso.2sls.control.lambda)
lasso.2sls.sensitivity.predict<-predict(lasso.2sls.
sensitivity, as.matrix(example.x.sensitivity), s=lasso.2sls.
sensitivity.lambda)

#Fit regression model using the predicted values#
lasso.2sls.ols<-lm(example$SCQ_DEN_1~lasso.2sls.negative.
predict+lasso.2sls.extraversion.predict+lasso.2sls.control.
predict+lasso.2sls.sensitivity.predict)

summary(lasso.2sls.ols)

Call:
lm(formula = example$SCQ_DEN_1 ~ lasso.2sls.negative.predict +
    lasso.2sls.extraversion.predict + lasso.2sls.control.
predict +
    lasso.2sls.sensitivity.predict)

Residuals:
    Min       1Q   Median       3Q      Max
-2.02018 -0.71370 -0.04079  0.56194  2.61567

Coefficients:
                                Estimate Std. Error t value Pr(>|t|)
(Intercept)                      2.99670    1.61242   1.859   0.0685 .
lasso.2sls.negative.predict      0.66274    0.30138   2.199   0.0321 *
lasso.2sls.extraversion.predict -0.04892    0.12824  -0.381   0.7043
lasso.2sls.control.predict      -0.10035    0.12610  -0.796   0.4296
lasso.2sls.sensitivity.predict  -0.15248    0.30364  -0.502   0.6175
---
Signif. codes:  0 '***' 0.001 '**' 0.01 '*' 0.05 '.' 0.1 ' ' 1
Residual standard error: 0.9449 on 55 degrees of freedom
Multiple R-squared: 0.09008, Adjusted R-squared: 0.0239
F-statistic: 1.361 on 4 and 55 DF, p-value: 0.2593
```

The results for the lasso penalized 2SLS estimators identified a statistically significant positive relationship between negative affect and coping. This finding differs from those obtained from the other estimators that we have examined in this section of the chapter. Taken together, these predictors account for approximately 9% of the variance in coping.

The ridge estimator can be applied with 2SLS in much the same fashion as was the lasso. The same basic steps are applied with the difference being that we set `alpha=0` for the ridge. The R commands for this purpose appear as follows:

```
#Ridge 2sls#
#Stage 1#
ridge.2sls.negative<-glmnet(as.matrix(example.x.negative),e
xample$ATS_FEAR_1,alpha=0, nlambda=100)
ridge.2sls.extraversion<-glmnet(as.matrix(example.x.extrave
rsion),example$ATS_SOCI_1,alpha=0, nlambda=100)
ridge.2sls.control<-glmnet(as.matrix(example.x.control),exa
mple$ATS_ATTC_1,alpha=0, nlambda=100)
ridge.2sls.sensitivity<-glmnet(as.matrix(example.x.sensitiv
ity),example$ATS_NPS_1,alpha=0, nlambda=100)

#Identify optimal tuning parameter values#
ridge.2sls.negative.cv<-cv.glmnet(as.matrix(example.x.negat
ive),example$ATS_FEAR_1)
ridge.2sls.extraversion.cv<-cv.glmnet(as.matrix(example.x.e
xtraversion),example$ATS_SOCI_1)
ridge.2sls.control.cv<-cv.glmnet(as.matrix(example.x.contro
l),example$ATS_ACTC_1)
ridge.2sls.sensitivity.cv<-cv.glmnet(as.matrix(example.x.se
nsitivity),example$ATS_NPS_1)

#Save optimal tuning parameter values#
ridge.2sls.negative.lambda<-ridge.2sls.negative.cv$lambda.min
ridge.2sls.extraversion.lambda<-ridge.2sls.extraversion.
cv$lambda.min
ridge.2sls.control.lambda<-ridge.2sls.control.cv$lambda.min
ridge.2sls.sensitivity.lambda<-ridge.2sls.sensitivity.
cv$lambda.min

#Obtain predicted values based on instruments#
ridge.2sls.negative.predict<-predict(ridge.2sls.negative,
as.matrix(example.x.negative), s=ridge.2sls.negative.
lambda)
ridge.2sls.extraversion.predict<-predict(ridge.2sls.
extraversion, as.matrix(example.x.extraversion),
s=ridge.2sls.extraversion.lambda)
ridge.2sls.control.predict<-predict(ridge.2sls.control,
as.matrix(example.x.control), s=ridge.2sls.control.lambda)
ridge.2sls.sensitivity.predict<-predict(ridge.2sls.
sensitivity, as.matrix(example.x.sensitivity),
s=ridge.2sls.sensitivity.lambda)

#Fit regression model using the predicted values#
ridge.2sls.ols<-lm(example$SCQ_DEN_1~ridge.2sls.negative.
predict+ridge.2sls.extraversion.predict+ridge.2sls.control.
predict+ridge.2sls.sensitivity.predict)
```

```
summary(ridge.2sls.ols)

Call:
lm(formula = example$SCQ_DEN_1 ~ ridge.2sls.negative.predict +
    ridge.2sls.extraversion.predict + ridge.2sls.control.
predict +
    ridge.2sls.sensitivity.predict)

Residuals:
    Min       1Q    Median       3Q       Max
-2.00739 -0.70063 -0.02935   0.52878   2.63927

Coefficients:
                                 Estimate Std. Error t value Pr(>|t|)
(Intercept)                       4.17216    2.20697   1.890   0.0640 .
ridge.2sls.negative.predict       0.66104    0.29808   2.218   0.0307 *
ridge.2sls.extraversion.predict  -0.05227    0.13685  -0.382   0.7040
ridge.2sls.control.predict       -0.39678    0.47168  -0.841   0.4039
ridge.2sls.sensitivity.predict   -0.15020    0.30983  -0.485   0.6298
---
Signif. codes:  0 '***' 0.001 '**' 0.01 '*' 0.05 '.' 0.1 ' ' 1

Residual standard error: 0.9442 on 55 degrees of freedom
Multiple R-squared: 0.09137, Adjusted R-squared: 0.02529
F-statistic: 1.383 on 4 and 55 DF, p-value: 0.2519
```

The elastic net estimator in the context of 2SLS can also be obtained using the approach outlined earlier. An additional issue that is unique to elastic net is that the optimal value of the α tuning parameter must be identified, in addition to that for λ. Thus, we would use the approach outlined in Chapter 3 for this purpose as a part of fitting estimator. In the following example, we use α=0.5 for pedagogical purposes. However, the reader should keep in mind that a thorough investigation of both λ and α would need to be carried out when using the elastic net.

```
#Net 2sls#
#Stage 1#
net.2sls.negative<-glmnet(as.matrix(example.x.negative),exa
mple$ATS_FEAR_1,alpha=0.5, nlambda=100)
net.2sls.extraversion<-glmnet(as.matrix(example.x.extravers
ion),example$ATS_SOCI_1,alpha=0.5, nlambda=100)
net.2sls.control<-glmnet(as.matrix(example.x.control),examp
le$ATS_ACTC_1,alpha=0.5, nlambda=100)
net.2sls.sensitivity<-glmnet(as.matrix(example.x.sensitivit
y),example$ATS_NPS_1,alpha=0.5, nlambda=100)

#Identify optimal tuning parameter values#
net.2sls.negative.cv<-cv.glmnet(as.matrix(example.x.negativ
e),example$ATS_FEAR_1)
net.2sls.extraversion.cv<-cv.glmnet(as.matrix(example.x.ext
raversion),example$ATS_SOCI_1)
```

```
net.2sls.control.cv<-cv.glmnet(as.matrix(example.x.control)
,example$ATS_ACTC_1)
net.2sls.sensitivity.cv<-cv.glmnet(as.matrix(example.x.sens
itivity),example$ATS_NPS_1)

#Save optimal tuning parameter values#
net.2sls.negative.lambda<-net.2sls.negative.cv$lambda.min
net.2sls.extraversion.lambda<-net.2sls.extraversion.
cv$lambda.min
net.2sls.control.lambda<-net.2sls.control.cv$lambda.min
net.2sls.sensitivity.lambda<-net.2sls.sensitivity.
cv$lambda.min

#Obtain predicted values based on instruments#
net.2sls.negative.predict<-predict(net.2sls.negative,
as.matrix(example.x.negative), s=net.2sls.negative.lambda)
net.2sls.extraversion.predict<-predict(net.2sls.
extraversion, as.matrix(example.x.extraversion),
s=net.2sls.extraversion.lambda)
net.2sls.control.predict<-predict(net.2sls.control,
as.matrix(example.x.control), s=net.2sls.control.lambda)
net.2sls.sensitivity.predict<-predict(net.2sls.sensitivity,
as.matrix(example.x.sensitivity), s=net.2sls.sensitivity.
lambda)

#Fit regression model using the predicted values#
net.2sls.ols<-lm(example$SCQ_DEN_1~net.2sls.negative.
predict+net.2sls.extraversion.predict+net.2sls.control.
predict+net.2sls.sensitivity.predict)
summary(net.2sls.ols)

Call:
lm(formula = example$SCQ_DEN_1 ~ net.2sls.negative.predict +
    net.2sls.extraversion.predict + net.2sls.control.predict +
    net.2sls.sensitivity.predict)

Residuals:
    Min       1Q    Median       3Q       Max
-2.01693 -0.71425 -0.04242  0.55954   2.61801

Coefficients:
                              Estimate Std. Error t value Pr(>|t|)
(Intercept)                    3.10660    1.59469   1.948   0.0565 .
net.2sls.negative.predict      0.63282    0.28855   2.193   0.0325 *
net.2sls.extraversion.predict -0.04989    0.12832  -0.389   0.6989
net.2sls.control.predict      -0.09993    0.12620  -0.792   0.4319
net.2sls.sensitivity.predict  -0.15074    0.30343  -0.497   0.6213
---
Signif. codes:  0 '***' 0.001 '**' 0.01 '*' 0.05 '.' 0.1 ' ' 1

Residual standard error: 0.9451 on 55 degrees of freedom
Multiple R-squared: 0.08968, Adjusted R-squared: 0.02348
F-statistic: 1.355 on 4 and 55 DF, p-value: 0.2616
```

Summary

In this chapter, we focused our attention on the application of regularization techniques to latent variable models, in particular factor analysis and SEM. We saw that data analysts have a large set of tools from which to choose when fitting these models. Indeed, other approaches that we did not touch on in this chapter, but which are also potentially useful for this purpose are approaches based on penalized likelihood (Huang, Chen, & Weng, 2017) and moderated nonlinear factor analysis (Bauer & Hussong, 2009). The approaches that we discussed in this chapter allow the researcher to fit standard latent variable models while at the same time taking advantage of regularization, which can be particularly useful for situations involving high-dimensional data structures. In many respects, regularized latent variable models, be they factor analysis or SEM, incorporate the same principles that were outlined in Chapter 2, and demonstrated in the preceding chapters. At the time of this writing, it is not possible to identify any one approach as being optimal or preferred vis-à-vis the others. Rather, it is recommended that the researcher use several of the methods outlined here and compare the results obtained from them in order to develop a full understanding of how regularization impacts the estimates when compared to the standard approach, and whether different approaches to penalizing the estimates yield different results. Such an analytic approach can be thought of as a type of sensitivity analysis. The more similar the various penalized estimator results are, the more confidence we can have in the generalizability of the results. On the other hand, if different penalty functions yield very different results, we would need to be more careful in our interpretation of the findings and generalization of the results.

In Chapter 8, we will conclude the book with a discussion of regularization techniques for multilevel models. We will see how multilevel models, in general, can be seen as direct extensions of single-level regression models. We will also learn that the techniques that have been featured throughout this book, including the lasso and ridge estimators, can be readily extended for use with models designed for nested data structures. Indeed, most of the principles that were first outlined in Chapter 2 and that have been foundational to the examples provided heretofore will also be features of the multilevel modeling that is described in the final chapter of the book.

8

Regularization Methods for Multilevel Models

In Chapters 3 and 4, we considered the standard linear model and its generalized linear model extension, which underlie standard linear regression, logistic regression, and regression models for other categorical outcomes. Each of these models rests on a variety of assumptions, with one common one across them being the assumption of independently distributed error terms for the individual observations within the sample. This assumption essentially means that there are no relationships among individuals in the sample for the dependent variable, *once the independent variables in the analysis are accounted for.* In many research scenarios, the sampling scheme can lead to correlated responses on the outcome among individuals. For example, a researcher interested in the impact of a new teaching method on student achievement might randomly select schools for placement in either a treatment or control group. If school A is placed into the treatment condition, all students within the school will also be in the treatment condition— this is a cluster-randomized design, in that the clusters and not the individuals are assigned to a specific group. Furthermore, it would be reasonable to assume that the school itself, above and beyond the treatment condition, would have an impact on the performance of the students. This impact would manifest itself as correlations in achievement test scores among individuals attending that school. Thus, if we were to use a regression model to assess the relationship between achievement test scores and treatment condition with such cluster sampled data, we would likely be violating the assumption of independent errors because a factor beyond treatment condition (in this case the school) would have an additional impact on the outcome variable.

We typically refer to the data structure described earlier as nested, meaning that individual data points at one level (e.g., student) appear in only one level of a higher-level variable such as school. Thus, students are nested within school. Such designs can be contrasted with a crossed data structure whereby individuals at the first level appear in multiple levels of the second variable. In our example, students might be crossed with after-school organizations if they are allowed to participate in more than one. For example, a given student might be on the basketball team as well as in the band. The focus of this chapter is on statistical analyses for dealing with such nested designs, and in particular approaches for regularized estimation in the multilevel context. We will start with an examination of multilevel models more generally, and then pivot to a discussion of how the regularization techniques that we

DOI: 10.1201/9780367809645-8

discussed in prior chapters can be easily extended to multilevel models. We will then see how penalized estimation can be easily applied to multilevel models for categorical outcome variables.

Multilevel Linear Regression

In the following section, we will review some of the core ideas that underlie multilevel linear models (MLMs) generally speaking. This is a very brief review of multilevel modeling designed to familiarize the reader with this framework but is not intended to be a completely thorough review of these methods. We will first focus on the difference between random and fixed effects, after which we will discuss the basics of parameter estimation, focusing on the two most commonly used methods, maximum likelihood and restricted maximum likelihood, and conclude with a review of assumptions underlying MLMs. We hope that after reading this section of the chapter, the reader will have sufficient technical background on MLMs to begin using the R software package for fitting MLMs of various types, and be ready to learn about the penalized extensions of these methods.

Random Intercept Model

As we transition to the MLM context, let's first revisit the basic simple linear regression model of equation:

$$y = \beta_0 + \beta_1 x + \varepsilon \tag{8.1}$$

Here, the dependent variable y is expressed as a function of an independent variable, x, multiplied by a slope coefficient, β_1, an intercept, β_0, and random variation from subject to subject, ε. We defined the intercept as the conditional mean of y when the value of x is 0. In the context of a single-level regression model such as this, there is one intercept that is common to all individuals in the population of interest. However, when individuals are clustered together in some fashion (e.g., within classrooms, schools, organizational units within a company), there will potentially be a separate intercept for each of these clusters; that is, there may be different means for the dependent variable for x=0 across the different clusters. In practice, assessing whether there are different means across the clusters is an empirical question, which we describe below. It should also be noted that in this discussion we are considering only the case where the intercept is cluster specific, but it is also possible for β_1 to vary by group as well, or even other coefficients from more complicated models.

Allowing for group-specific intercepts and slopes leads to the following notation commonly used for the level 1 (micro-level) model in multilevel modeling:

$$y_{ij} = \beta_{0j} + \beta_{1j}x + \varepsilon_{ij} \tag{8.2}$$

The subscripts *ij* refer to the *i*th individual in the *j*th cluster.

As we continue our discussion of multilevel modeling notation and structure we will begin with the most basic multilevel model: predicting the outcome from just an intercept which we will allow to vary randomly for each group.

$$y_{ij} = \beta_{0j} + \varepsilon_{ij} \tag{8.3}$$

Allowing the intercept to differ across clusters, as in equation 8.3, leads to the random intercept which we express as

$$\beta_{0j} = \gamma_{00} + U_{0j} \tag{8.4}$$

In this framework, γ_{00} represents an average or general intercept value that holds across clusters, whereas U_{0j} is a group specific effect on the intercept. We can think of γ_{00} as a fixed effect because it remains constant across all clusters, and U_{0j} is a random effect because it varies from cluster to cluster. Therefore, for an MLM we are interested not only in some general mean value for *y* when *x* is 0 for all individuals in the population (γ_{00}), but also the in deviation between the overall mean and the cluster-specific effects for the intercept (U_{0j}). If we go on to assume that the clusters are a random sample from the population of all such clusters, then we can treat U_{0j} as a residual effect for y_{ij}, very similar to how we think of ε. In that case, U_{0j} is assumed to be drawn randomly from a population with a mean of 0 (recall U_{0j} is a deviation from the fixed effect) and a variance, τ^2. Furthermore, we assume that τ^2 and σ^2, the variance of ε, are uncorrelated. In addition, τ^2 can also be viewed as the impact of the cluster on the dependent variable, and therefore testing it for statistical significance is equivalent to testing the null hypothesis that cluster (e.g., school) has no impact on the dependent variable. If we substitute the two components of the random intercept into the regression model, we get

$$y = \gamma_{00} + U_{0j} + \beta_1 x + \varepsilon \tag{8.5}$$

Often in MLM, we begin our analysis of a dataset with this simple random intercept model, known as the null model, which takes the form

$$y_{ij} = \gamma_{00} + U_{0j} + \varepsilon_{ij} \tag{8.6}$$

While the null model does not provide information regarding the impact of specific independent variables on the dependent, it does yield important information regarding how variation in *y* is partitioned between variance among the individuals σ^2 and variance among the clusters τ^2. The total variance of y is simply the sum of σ^2 and τ^2. In addition, as we have already seen, these values can be used to estimate ρ_I. The null model, as will be seen in later sections, is also used as a baseline for model building and comparison.

Random Coefficients Model

It is a simple matter to expand the random intercept model in (8.5) to accommodate one or more independent predictor variables. As an example, if we add a single predictor (x_{ij}) at the individual level (level-1) to the model, we obtain

$$y_{ij} = \gamma_{00} + \gamma_{10}x_{ij} + U_{0j} + \varepsilon_{ij} \tag{8.7}$$

This model can also be expressed in two separate levels as:

$$\text{Level 1: } y_{ij} = \beta_{0j} + \beta_{1j}x + \varepsilon_{ij} \tag{8.8}$$

$$\text{Level 2: } \beta_{0j} = \gamma_{00} + U_{0j} \tag{8.9}$$

$$\beta_{1j} = \gamma_{10} \tag{8.10}$$

This model now includes the predictor and the slope relating it to the dependent variable, γ_{10}, which we acknowledge as being at level-1 by the subscript 10. We interpret γ_{10} in the same way that we did β_1 in the linear regression model; i.e., a measure of the impact on y of a 1 unit change in x. In addition, we can estimate ρ_I exactly as before, though now it reflects the correlation between individuals from the same cluster after controlling for the independent variable, x. In this model, both γ_{10} and γ_{00} are fixed effects, while σ^2 and τ^2 remain random.

One implication of this model is that the dependent variable is impacted by variation among individuals (σ^2), variation among clusters (τ^2), an overall mean common to all clusters (γ_{00}), and the impact of the independent variable as measured by γ_{10}, which is also common to all clusters. In practice, there is no reason that the impact of x on y would need to be common for all clusters, however. In other words, it is entirely possible that rather than having a single γ_{10} common to all clusters, there is actually a unique effect for the cluster of $\gamma_{10} + U_{1j}$, where γ_{10} is the average relationship of x with y across clusters, and U_{1j} is the cluster-specific variation of the relationship between the two variables. This cluster-specific effect is assumed to have a mean of 0 and to vary randomly around γ_{10}. The random slopes model is

$$y_{ij} = \gamma_{00} + \gamma_{10}x_{ij} + U_{0j} + U_{1j}x_{ij} + \varepsilon_{ij} \tag{8.11}$$

Written in this way, we have separated the model into its fixed $(\gamma_{00} + \gamma_{10}x_{ij})$ and random $(U_{0j} + U_{1j}x_{ij} + \varepsilon_{ij})$ components. Model (8.11) simply states that there is an interaction between cluster and x, such that the relationship of x and y is not constant across clusters.

Heretofore we have discussed only one source of between-group variation, which we have expressed as τ^2, and which is the variation among clusters in the intercept. However, model (8.11) adds a second such source of between-group variance in the form of U_{1j}, which is cluster variation on the slope relating the independent and dependent variables. In order to differentiate between these two sources of between-group variance, we now denote the variance of U_{0j} as τ_0^2 and the variance of U_{1j} as τ_1^2. Furthermore, within clusters we expect U_{1j} and U_{0j} to have a covariance of τ_{01}. However, across different clusters these terms should be independent of one another, and in all cases, it is assumed that ε remains independent of all other model terms. In practice, if we find that τ_1^2 is not 0, we must be careful in describing the relationship between the independent and dependent variables, as it is not the same for all clusters. We will revisit this idea in subsequent chapters. For the moment, however, it is most important to recognize that variation in the dependent variable, y, can be explained by several sources, some fixed and others random. In practice, we will most likely be interested in estimating all of these sources of variability in a single model.

Regularized Multilevel Regression Model

As we have discussed throughout this book, in some research contexts the number of variables that can be measured (p) approaches, or even exceeds the number of individuals on whom such measurements can be made (N), yielding high-dimensional data. We have seen that in such cases, standard statistical models often do not work well, yielding biased standard errors for the model parameter estimates (Bühlmann & van de Geer, 2011). These biased standard errors in turn can lead to inaccurate Type I error and power rates for inferences made about these parameters. High dimensionality can also result in parameter estimation bias due to the presence of collinearity (Fox, 2016). Finally, when p exceeds N, it may not be possible to obtain parameter estimates at all using standard estimators.

In Chapter 2, we described in some detail a set of penalized estimators that have been designed for use in cases when standard estimators (such as maximum likelihood) may not work particularly well. These methods can be easily applied to the multilevel modeling context, as described earlier. Schelldorfer, Bühlmann, and van de Geer (2011) described an extension of the lasso estimator that we have used throughout this book for use with multilevel models. The multilevel lasso (MLL) utilizes the lasso penalty function, with additional terms to account for the variance components associated with multilevel models. The MLL estimator minimizes the following function:

$$Q_\lambda\left(\beta,\tau^2,\sigma^2\right) := \frac{1}{2}ln|V| + \frac{1}{2}(y_i - \hat{y}_i)^{\prime V^{-1}(y_i - \hat{y}_i)} + \lambda\sum_{j=1}^{p}\left\lfloor \hat{\beta}_j \right\rfloor \qquad (8.12)$$

where

τ^2 = Between cluster variance at level-2

σ^2 = Within cluster variance at level-1

V = Covariance matrix

From equation (8.12), we can see that model parameter estimates are obtained with respect to penalization of level-1 coefficients and otherwise works similarly to the single level lasso estimator. In order to conduct inference for the MLL model parameters, standard errors must be estimated. However, the MLL algorithm currently does not provide standard error estimates, meaning that inference is not possible. Therefore, interpretation of results from analyses using MLL will focus on which coefficients are not shrunken to 0, as we will see in the example below.

Fitting the Multilevel Lasso in R

In order to demonstrate the use of the multilevel lasso model, we will use the classroomStudy data that is part of the lmmlasso R library. This library is not currently available using the standard downloading and installation approach that we have used previously in this book. However, it remains, in my opinion, the optimal package for fitting penalized estimators in the multilevel modeling framework. Therefore, we will use it in this chapter. It can be downloaded from https://cran.r-project.org/web/packages/lmmlasso/index.html. Once the compressed file is downloaded, it can be installed using the following command:

```
install.packages("c:\\research\\regularization book\\
lmmlasso_0.1-2.tar.gz", repos = NULL, type="source")
```

The reader will need to change the address to match the location of the file on their computer.

In this example, the gain in student math achievement scores will serve as the dependent variable, with the independent variables including student gender, ethnic minority status, type of math instruction, students' socioeconomic status, years of teaching experience, and time spent in math preparation. The level-2 effect was classroom. The level-1 sample size was 156 students, with level-2 having 44 classrooms.

We will first fit this model using the standard multilevel model, using lme4. The syntax requires the dependent variable (y) to be followed by ~ and then the random effect. We can obtain the parameter estimates using the summary command, as well as 95% confidence intervals for all model parameter estimates using the percentile bootstrap.

```
library(lme4)
library(lmmlasso)
data(classroomStudy)

   Model8.0<-lmer(y~(1|grp), data=classroomStudy)

   summary(Model8.0)

   Linear mixed model fit by REML ['lmerMod']
   Formula: y ~ (1 | classroomStudy$grp)
      Data: classroomStudy

   REML criterion at convergence: 437.5

   Scaled residuals:
        Min      1Q  Median      3Q     Max
   -2.0789 -0.6311 -0.1229  0.5766  3.2235

   Random effects:
    Groups              Name          Variance Std.Dev.
    classroomStudy$grp (Intercept) 0.2632   0.5131
    Residual                         0.7792   0.8827
   Number of obs: 156, groups:  classroomStudy$grp, 44

   Fixed effects:
                Estimate Std. Error t value
   (Intercept)  0.04269    0.11040   0.387

   confint(Model8.0, method=c("boot"), boot.type=c("perc"))
   Computing bootstrap confidence intervals . . .

   2 message(s): boundary (singular) fit: see ?isSingular

                  2.5 %      97.5 %
   .sig01      0.2213364 0.7133444
   .sigma      0.7608479 0.9999528
   (Intercept) -0.1809469 0.2680129
```

The first model that we fit includes only the clustering variable (grp), which allows us to estimate the Intraclass correlation (ICC) of the math test scores. The ICC is a measure of the proportion of variance in the outcome variable that is associated with the clustering variable. Based on the confidence intervals for the parameter estimates, we conclude that there was a statistically significant amount of variability in the math score variation due to classroom

(0.22, 0.71) and individual examinee (0.76, 1.00). The confidence interval for the intercept term included 0 (−0.18, 0.27), indicating that the mean math score was not significantly different from 0. This is expected because these scores are standardized with a mean of 0 and variance of 1. Using these results, we can calculate the ICC as the ratio of the classroom variance in math scores to the total variance (classroom + student).

$$ICC = \frac{0.26}{0.26 + 0.78} = 0.25$$

Thus, it appears that approximately 25% of the variance in math test scores is associated with the classroom.

Next, we can fit a full model including the predictors for math achievement scores, while accounting for the variance associated with classroom. The standard estimator using lmer looks much as that for the null model, with the inclusion of the predictor variables, as below. Note that in this example, the predictor variables are contained in the matrix x within the clasrrom-Study dataframe. The first column of this matrix is a column of 1's, which is necessary for the lmmlasso function, but which is not necessary for lmer.

```
Model8.1<-lmer(y~X[,2:7]+(1|classroomStudy$grp),
data=classroomStudy)

summary(Model8.1)

Linear mixed model fit by REML ['lmerMod']
Formula: y ~ X[, 2:7] + (1 | classroomStudy$grp)
   Data: classroomStudy

REML criterion at convergence: 375

Scaled residuals:
    Min      1Q  Median      3Q     Max
-2.3869 -0.6120 -0.1356  0.6665  2.2173

Random effects:
 Groups              Name         Variance Std.Dev.
 classroomStudy$grp (Intercept) 0.1583   0.3979
 Residual                        0.4745   0.6888

Number of obs: 156, groups:  classroomStudy$grp, 44
Fixed effects:
                  Estimate Std. Error t value
(Intercept)        0.02599    0.08611   0.302
X[, 2:7]sex       -0.07690    0.06046  -1.272
X[, 2:7]minority  -0.02015    0.06803  -0.296
X[, 2:7]mathkind  -0.63959    0.06356 -10.063
X[, 2:7]ses        0.05329    0.06119   0.871
X[, 2:7]yearstea  -0.01861    0.09656  -0.193
X[, 2:7]mathprep  -0.06041    0.10045  -0.601
```

```
Correlation of Fixed Effects:
            (Intr) X[,2:7]sx X[,2:7]mn X[,2:7]mthk X[,2:7]ss X[,2:7]y
X[, 2:7]sex -0.013
X[,2:7]mnrt -0.022   0.133
X[,2:7]mthk  0.024  -0.059      -0.042
X[, 2:7]ses  0.031  -0.046      -0.031       -0.116
X[,2:7]yrst -0.044   0.013       0.040       -0.068      -0.088
X[,2:7]mthp  0.059  -0.013       0.011        0.093      -0.009   -0.419
> confint(Model8.1, method=c("boot"), boot.type=c("perc"))
Computing bootstrap confidence intervals . . .

2 message(s): boundary (singular) fit: see ?isSingular
1 warning(s): Model failed to converge with max|grad| =
0.00201395 (tol = 0.002, component 1)
                              2.5 %        97.5 %
.sig01                   0.21567420    0.58583411
.sigma                   0.59450602    0.77616229
(Intercept)             -0.14960147    0.20199408
X[, 2:7]sex             -0.20389611    0.03690808
X[, 2:7]minority        -0.15308296    0.11245640
X[, 2:7]mathkind        -0.76767686   -0.51276183
X[, 2:7]ses             -0.07261175    0.16294782
X[, 2:7]yearstea        -0.20489393    0.18146114
X[, 2:7]mathprep        -0.29078249    0.14486664
```

From these results, we would conclude that the only statistically significant effect was type of math instruction, with a coefficient of −0.64. It was the only variable for which the confidence interval did not include 0 (−0.77, −0.51).

Next, let's fit the same model using the multilevel lasso. In order to do this, we will need to install and then load the lmmlasso package in R, which we described earlier. The R commands to fit the model, and to obtain a summary of the output appear as follows:

```
Model8.2 <-lmmlasso(x=classroomStudy$X,y=classroomStudy$y,z
=classroomStudy$Z,grp=classroomStudy$grp,lambda=15,pdMat="p
dIdent")
summary(Model8.2)
```

There are several things to note when using lmmlasso. First, the fixed effects variables are collected in the matrix X, which we have already commented on when discussing the use of lmer with the classroomStudy data. When using lmmlasso, the fixed effects will always need to be in such a matrix. Second, the object classroomStudy$Z is a column of 1's in this case, indicating that we have only a random slope. If we were also fitting random slopes for one or more independent variables, then those variables would need to each have a column in the Z matrix. The level-2 clustering variable, grp, appears after the grp subcommand. We must set the value of lambda, which in this case is 15. We will try other values, and compare model fit using AIC and BIC, selecting the lambda that minimizes these values. Finally, the pdMat subcommand defines the covariance structure for the random effect, in this case setting it to be the identity matrix.

The output for Model8.2 appears as follows:

```
Model fitted by ML for lambda = 15 :
      AIC        BIC     logLik  deviance objective
    363.9      379.1     -176.9     353.9      186.0

Random effects: pdIdent
            Variance   Std.Dev.
Intercept 0.1227391 0.3503414
Residual  0.4769248 0.6905974

Fixed effects:
|active set|= 3
                Estimate
(Intercept)  0.02065969 (n)
sex         -0.02376197
mathkind    -0.57887176

Number of iterations: 4
```

Using a lambda of 15, only two fixed effects had non-zero coefficients, sex and mathkind. Also, notice that the coefficients for these effects are smaller than were those estimated using the standard multilevel model, reflecting the shrinkage associated with the lasso. Let's now fit a model with a lambda of 10.

```
Model8.3 <-lmmlasso(x=classroomStudy$X,y=classroomStudy$y,z
=classroomStudy$Z,grp=classroomStudy$grp,lambda=10,pdMat="p
dIdent")
summary(Model8.3)
Model fitted by ML for lambda = 10 :
      AIC        BIC     logLik  deviance objective
    364.8      383.1     -176.4     352.8      182.9

Random effects: pdIdent
            Variance   Std.Dev.
Intercept 0.1211003 0.3479947
Residual  0.4742756 0.6886767

Fixed effects:
|active set|= 4
                Estimate
(Intercept)  0.02039525 (n)
sex         -0.04084089
mathkind    -0.59798409
ses          0.00546273

Number of iterations: 5
```

We can see that there is less shrinkage in the fixed effects coefficients, which we expect given that the penalty term for this model is smaller. However, also notice that the AIC and BIC values are larger for Model8.3 than was the case for Model8.2, leading us to conclude that we may need a larger penalty term. Next, let's set lambda at 20 and see what happens to the information indices.

```
Model8.4 <-lmmlasso(x=classroomStudy$X,y=classroomStudy$y,z
=classroomStudy$Z,grp=classroomStudy$grp,lambda=20,pdMat="p
dIdent")
summary(Model8.4)
Model fitted by ML for lambda = 20 :
     AIC        BIC     logLik   deviance  objective
    365.1      380.4     -177.6     355.1      188.9

Random effects: pdIdent
            Variance   Std.Dev.
Intercept  0.1247968  0.3532659
Residual   0.4802174  0.6929772

Fixed effects:
|active set|= 3
                  Estimate
(Intercept)    0.021169801  (n)
sex           -0.006581567
mathkind      -0.560115122

Number of iterations: 5
```

The fixed effects parameter estimate shrinkage is greater when lambda is 20, as we would anticipate. The AIC and BIC values are larger than those for a lambda of 15, indicating that the latter is a preferable value in terms of model fit. We could repeat this sequence of model fitting and comparison of information indices for various values of lambda in order to obtain what we believe to be the optimal setting. Given what we have already seen here, it is most likely that this optimal lambda value is close to 15, and so we would concentrate our exploration in that neighborhood.

Multilevel Logistic Regression Model

In order to introduce multilevel generalized linear models (MGLMs) for dichotomous outcomes, let us consider the following example. A researcher has collected testing data indicating whether 9,316 students have passed a state mathematics assessment, along with several measures of mathematics aptitude that were measured prior to administration of the achievement test.

She is interested in whether there exists a relationship between the score on number sense aptitude and the likelihood that a student will achieve a passing score on the mathematics achievement test, for which all examinees are categorized as either passing (1) or failing (0). Given that the outcome variable is dichotomous, we could use the binary logistic regression method introduced in Chapter 4. However, students in this sample are clustered by school, and therefore, we will need to appropriately account for this multilevel data structure in our regression analysis.

Random Intercept Logistic Regression

In order to fit the random intercept logistic regression model in R, we will use the glmer function for fitting GLiMs that is associated with the lme4 library. The algorithm underlying glmer relies on an adaptive Gauss-Hermite likelihood approximation (Liu & Pierce, 1994) to fit the model to the data. Given the applied focus of this book, we will not devote time to the technical specifications of this method for fitting the model to the data. However, the interested reader is referred to the Liu and Pierce work for a description of this method. The R command for fitting the model and obtaining the summary statistics appear below, following the call to the lme4 library and the attaching of the file containing the data. Note that in this initial analysis, we have a fixed effect for the intercept and the slope of the independent variable numsense, but we only allow a random intercept, thereby assuming that the relationship between the number sense score and the likelihood of achieving a passing score on the state math assessment (score2) is fixed across schools; i.e., the relationship of numsense with score2 does not vary from one school to another.

```
library(lme4)

model8.5<-glmer(score2~numsense+L_Free+L_Reduced+Salary+
ISTEPM+specialed+female+(1|school),family=binomial,
data=mathfinal.nomiss, na.action=na.omit)

summary(model8.5)

Generalized linear mixed model fit by maximum likelihood
(Laplace Approximation) [
glmerMod]
  Family: binomial  ( logit )
Formula: score2 ~ numsense + L_Free + L_Reduced + Salary +
ISTEPM + specialed +
    female + (1 | school)
   Data: mathfinal.nomiss
```

```
       AIC       BIC    logLik deviance df.resid
    7033.2    7094.6   -3507.6    7015.2      6801

Scaled residuals:
     Min      1Q   Median       3Q      Max
 -4.9439 -0.6768   0.2712   0.6251   3.7613

Random effects:
 Groups Name           Variance Std.Dev.
 school (Intercept) 0.1753      0.4187
Number of obs: 6810, groups:   school, 32

Fixed effects:
                Estimate Std. Error z value Pr(>|z|)
(Intercept) -1.137e+01   5.066e-01 -22.444  < 2e-16 ***
numsense     6.305e-02   1.769e-03  35.632  < 2e-16 ***
L_Free      -7.019e-03   3.126e-03  -2.245   0.0247 *
L_Reduced   -8.723e-03   2.080e-02  -0.419   0.6750
Salary      -2.455e-05   8.554e-06  -2.870   0.0041 **
ISTEPM       1.926e-01   7.646e-02   2.519   0.0118 *
specialed   -9.254e-01   1.784e-01  -5.187 2.13e-07 ***
female      -4.078e-02   5.904e-02  -0.691   0.4897
---
Signif  codes: 0 '***' 0.001 '**' 0.01 '*' 0.05 '.' 0.1 ' ' 1

Correlation of Fixed Effects:
            (Intr) numsns L_Free L_Rdcd Salary ISTEPM specld
numsense    -0.652
L_Free      -0.242  0.037
L_Reduced   -0.239 -0.014 -0.046
Salary      -0.461 -0.109 -0.189 -0.029
ISTEPM      -0.247  0.026  0.029 -0.102 -0.038
specialed   -0.041 -0.017  0.003 -0.027  0.056  0.037
female      -0.058 -0.002 -0.006 -0.005  0.006 -0.004  0.064
```

The function call is similar to what we saw with linear models above. In terms of interpretation of the results, we first examine the variability in intercepts from school to school. This variation is presented as both the variance and standard deviation of the U_{0j} terms from Chapter 2 (τ_0^2), which are 0.2888 and 0.5374, respectively, for this example. The modal value of the intercept across schools is –11.659653. With regard to the fixed effect, the slope of numsense, we see that higher scores are associated with a greater likelihood of passing the state math assessment, with the slope being 0.059177 ($p < .05$). (Remember that R models the larger value of the outcome in the numerator of the logit, and in this case, passing was coded as 1 whereas failing was coded as 0.) The standard error, test statistic, and p-value appear in the next three columns. The results are statistically significant, ($p<0.001$), leading to the conclusion that overall, number sense scores are positively related to the likelihood of a student achieving a passing score on the assessment. Finally,

we see that the correlation between the slope and intercept is strongly negative (−0.955). Given that this is an estimate of the relationship between two
fixed effects, we are not particularly interested in it. Information about the
residuals appear at the very end of the output.

Using the `glmer` function, we can also fit a logistic regression model
with random coefficients. In the following example, we will take model8.5
and include random coefficient terms for the number sense (numsense)
and standardized math test score (ISTEPM). This is done by including the
numsense+ISTEPM|schoolrandom effect in the model, as we see below.

```
model8.6<-glmer(score2~numsense+L_Free+L_Reduced+Salary+IST
EPM+specialed+female+(numsense+ISTEPM|school),family=binomi
al, data=mathfinal.nomiss, na.action=na.omit)

summary(model8.6)

Generalized linear mixed model fit by maximum likelihood
(Laplace Approximation) [
glmerMod]
 Family: binomial  ( logit )
Formula: score2 ~ numsense + L_Free + L_Reduced + Salary +
ISTEPM + specialed +
    female + (numsense + ISTEPM | school)
   Data: mathfinal.nomiss

    AIC       BIC    logLik deviance df.resid
 7000.0    7095.6   -3486.0   6972.0      6796

Scaled residuals:
    Min      1Q  Median      3Q     Max
-4.6862 -0.6576  0.2666  0.6199  3.7377

Random effects:
 Groups Name        Variance  Std.Dev. Corr
 school (Intercept) 1.958e+01 4.42479
        numsense    3.662e-04 0.01914  -1.00
        ISTEPM      1.764e-02 0.13283  -0.11  0.11
Number of obs: 6810, groups:  school, 32

Fixed effects:
              Estimate Std. Error z value Pr(>|z|)
(Intercept) -1.177e+01  9.161e-01 -12.853  < 2e-16 ***
numsense     6.555e-02  3.928e-03  16.689  < 2e-16 ***
L_Free      -3.996e-03  2.793e-03  -1.431 0.152452
L_Reduced   -1.729e-02  1.631e-02  -1.060 0.289235
Salary      -2.708e-05  6.964e-06  -3.888 0.000101 ***
ISTEPM       1.091e-01  7.771e-02   1.403 0.160538
specialed   -9.115e-01  1.780e-01  -5.122 3.03e-07 ***
female      -3.758e-02  5.928e-02  -0.634 0.526108
---
Signif. codes: 0 '***' 0.001 '**' 0.01 '*' 0.05 '.' 0.1 ' ' 1
```

```
Correlation of Fixed Effects:
            (Intr) numsns L_Free L_Rdcd Salary ISTEPM specld
numsense   -0.923
L_Free     -0.086 -0.042
L_Reduced  -0.100 -0.078  0.067
Salary     -0.202 -0.081 -0.175  0.081
ISTEPM     -0.158  0.037 -0.029 -0.027  0.033
specialed  -0.025 -0.008  0.014 -0.039  0.075  0.031
female     -0.035  0.005 -0.022  0.000 -0.003  0.006  0.066
fit warnings:
Some predictor variables are on very different scales:
consider rescaling
convergence code: 0
boundary (singular) fit: see ?isSingular
failure to converge in 10000 evaluations
```

First, we should note that this model did not converge, meaning that we cannot confidently interpret the model parameter estimates. However, we do need to point out the relevant information that can be obtained from the random slopes model. The new pieces of output are the variance estimates for numsense and ISTEPM, which reflect the amount of variance across schools in the relationships between each of these variables and the outcome variable. The variance estimates were −.00037 for numsense and 0.1764 for ISTEPM. In addition, the correlation between the two random effects was −0.11, indicating a weak negative relationship in the amount of within-school variance for numsense and ISTEPM.

Regularized Multilevel Logistic Regression

We will start our investigation of the lasso logistic regression model by fitting a model with $\lambda=1$, using the glmmLasso function from the glmmLasso library. First, we load the library, and then specify the model and saving the results in the object model8.7a. The model includes the same dependent variable and fixed effects as for model8.5. The random effect is specified as rnd = list(school.f=~1). Next, we set the value of λ to 1, indicate the distributional family and link function for the dependent variable, and finally indicate which dataset should be used. We then request summary results for the model, with the resulting output appearing as follows:

```
library(glmmLasso)

model8.7a <- glmmLasso(score2~numsense+L_Free+L_Reduced+Salar
y+ISTEPM+specialed+female, rnd = list(school.f=~1), lambda=1,
family=binomial(link="logit"), data = mathfinal.nomiss)
```

```
summary(model8.7a)

Call:
glmmLasso(fix = score2 ~ numsense + L_Free + L_Reduced + Salary +
    ISTEPM + specialed + female, rnd = list(school.f = ~1),
data = mathfinal.nomiss,
    lambda = 1, family = binomial(link = "logit"))

Fixed Effects:

Coefficients:
                Estimate StdErr z.value p.value
(Intercept) -1.1124e+01      NA      NA      NA
numsense     6.1627e-02      NA      NA      NA
L_Free      -6.6616e-03      NA      NA      NA
L_Reduced    0.0000e+00      NA      NA      NA
Salary      -2.5371e-05      NA      NA      NA
ISTEPM       1.8027e-01      NA      NA      NA
specialed   -9.5421e-01      NA      NA      NA
female      -3.6945e-02      NA      NA      NA

Random Effects:

StdDev:
           school.f
school.f 0.3640347

model8.7a$aic
          [,1]
[1,]  7026.183
model8.7a$bic
          [,1]
[1,]  7225.027
```

The fixed effects estimates appear in the coefficients table. Note that the StdErr, z.value, and p. value columns all contain NA. An exploration of online discussion forums and help pages during the late summer of 2021 indicates that this is an endemic issue for glmmLasso. It is unclear why the standard errors are not estimated. We will offer an alternative approach to inference for the model parameter estimates in the form of a confidence interval. With respect to the estimates themselves, the value for L _ Reduced was regularized to 0, and the others were similar in value (though usually somewhat smaller) than those from the standard estimator. The standard deviation estimate for the school intercept random effect was slightly smaller than that of the maximum likelihood estimator (0.36 versus 0.42). The variance estimate for school is calculated as $0.3640347^2 = 0.133$, which compares to 0.175 for the standard estimator. Finally, we will want to use the AIC and BIC values to compare the fit of the models with different values for λ, with the model having the smallest values being selected as optimal.

Now let's fit the model with a λ value of 5.

```
model8.7b <- glmmLasso(score2~numsense+L_Free+L_Reduced+
Salary+ISTEPM+specialed+female, rnd = list(school.f=~1),
lambda=5, family=binomial(link="logit"), data = mathfinal.
nomiss)

summary(model8.7b)

Call:
glmmLasso(fix = score2 ~ numsense + L_Free + L_Reduced +
Salary +
    ISTEPM + specialed + female, rnd = list(school.f = ~1),
data = mathfinal.nomiss,
    lambda = 5, family = binomial(link = "logit"))

Fixed Effects:

Coefficients:
              Estimate StdErr z.value p.value
(Intercept) -1.1089e+01     NA      NA      NA
numsense     6.1264e-02     NA      NA      NA
L_Free      -6.3927e-03     NA      NA      NA
L_Reduced    0.0000e+00     NA      NA      NA
Salary      -2.4462e-05     NA      NA      NA
ISTEPM       1.7181e-01     NA      NA      NA
specialed   -9.3004e-01     NA      NA      NA
female      -2.9540e-02     NA      NA      NA

Random Effects:

StdDev:
            school.f
school.f 0.3592346

model8.7b$aic
         [,1]
[1,] 7027.458
model8.7b$bic
         [,1]
[1,] 7225.622
```

Our primary interest at this point is the comparison of AIC and BIC values, which are smaller for the $\lambda=1$ model than for $\lambda=5$.

Finally, we will try $\lambda=10$.

```
model8.7c <- glmmLasso(score2~numsense+L_Free+L_Reduced+
Salary+ISTEPM+specialed+female, rnd = list(school.f=~1),
lambda=10, family=binomial(link="logit"), data = mathfinal.
nomiss)

summary(model8.7c)

Call:
glmmLasso(fix = score2 ~ numsense + L_Free + L_Reduced +
```

```
Salary +
    ISTEPM + specialed + female, rnd = list(school.f = ~1),
data = mathfinal.nomiss,
    lambda = 10, family = binomial(link = "logit"))

Fixed Effects:

Coefficients:
              Estimate StdErr z.value p.value
(Intercept) -1.1041e+01    NA      NA      NA
numsense     6.0798e-02    NA      NA      NA
L_Free      -6.0687e-03    NA      NA      NA
L_Reduced    0.0000e+00    NA      NA      NA
Salary      -2.3395e-05    NA      NA      NA
ISTEPM       1.6175e-01    NA      NA      NA
specialed   -9.0088e-01    NA      NA      NA
female      -2.0329e-02    NA      NA      NA

Random Effects:

StdDev:
            school.f
school.f 0.3520709

model8.7c$aic
         [,1]
[1,] 7029.755

> model8.7c$bic
         [,1]
[1,] 7226.835
```

Based on the AIC and BIC values, of these three λ values, we would select 1. Now that we have selected a model, it would be helpful to have some notion regarding which of the model parameters are likely to differ from 0 in the population. As we have already discussed, currently the output from glmmLasso doesn't provide information on inference for the model parameters. However, we can use a bootstrap approach based on the boot library in R to calculate 95% confidence intervals for each model parameter estimate. A function to bootstrap resample this model appears at the book website (www.routledge.com/9780367408787), with documentation internal to the file.

Once we have submitted the bootstrap file to R, we then run the following command, saving the resamples in the object test.result. In the function call, we indicate which dataset to use, the name of the bootstrap function (glmmlasso.boot), and the number of bootstrap samples (100). This function takes quite some time to execute, which is why the value was set to 100 for this example. In practice, data analysts with faster computers and sufficient time should consider using a larger number, such as 500 or 1000.

```
test.result<-boot(data=mathfinal.nomiss,
statistic=glmmlasso.boot, R=100)
```

We can then obtain a confidence interval using the `boot.ci` function. In the function call, we indicate the resampled object, the type of confidence interval (here we request the percentile bootstrap), and the column of interest. The columns represent the model parameter estimates, with the order matching that in the output above (`intercept, numsense, L _ Free, ..., female`). Thus `index=1` refers to the intercept, `index=2` to `numsense`, and so on.

```
boot.ci(test.result, type="perc", index=1)
BOOTSTRAP CONFIDENCE INTERVAL CALCULATIONS
Based on 100 bootstrap replicates

CALL :
boot.ci(boot.out = test.result, type = "perc", index = 1)
#INTERCEPT#

Intervals :
Level      Percentile
95%    (-11.82, -10.81 )
Calculations and Intervals on Original Scale
boot.ci(test.result, type="perc", index=2)
BOOTSTRAP CONFIDENCE INTERVAL CALCULATIONS
Based on 100 bootstrap replicates

CALL :
boot.ci(boot.out = test.result, type = "perc", index = 2)
#NUMSENSE#

Intervals :
Level      Percentile
95%    ( 0.0598,  0.0641 )
Calculations and Intervals on Original Scale
boot.ci(test.result, type="perc", index=3)
BOOTSTRAP CONFIDENCE INTERVAL CALCULATIONS
Based on 100 bootstrap replicates

CALL :
boot.ci(boot.out = test.result, type = "perc", index = 3)
#L_FREE#

Intervals :
Level      Percentile
95%    (-0.0098, -0.0058 )
Calculations and Intervals on Original Scale
boot.ci(test.result, type="perc", index=4)
BOOTSTRAP CONFIDENCE INTERVAL CALCULATIONS
Based on 100 bootstrap replicates

CALL :
boot.ci(boot.out = test.result, type = "perc", index = 4)
#L_REDUCED#

Intervals :
Level      Percentile
```

```
95%    ( 0,   0 )
Calculations and Intervals on Original Scale

boot.ci(test.result, type="perc", index=5)
BOOTSTRAP CONFIDENCE INTERVAL CALCULATIONS
Based on 100 bootstrap replicates

CALL :
boot.ci(boot.out = test.result, type = "perc", index = 5)
#SALARY#

Intervals :
Level      Percentile
95%    ( 0,   0 )
Calculations and Intervals on Original Scale
boot.ci(test.result, type="perc", index=6)
BOOTSTRAP CONFIDENCE INTERVAL CALCULATIONS
Based on 100 bootstrap replicates

CALL :
boot.ci(boot.out = test.result, type = "perc", index = 6) #ISTEPM#
Intervals :
Level      Percentile
95%    ( 0.1417,  0.2302 )
Calculations and Intervals on Original Scale
boot.ci(test.result, type="perc", index=7)
BOOTSTRAP CONFIDENCE INTERVAL CALCULATIONS
Based on 100 bootstrap replicates

CALL :
boot.ci(boot.out = test.result, type = "perc", index = 7)
#SPECIALED#

Intervals :
Level      Percentile
95%    (-1.1957, -0.7306 )
BOOTSTRAP CONFIDENCE INTERVAL CALCULATIONS
Based on 100 bootstrap replicates

boot.ci(test.result, type="perc", index=8) #FEMALE#
BOOTSTRAP CONFIDENCE INTERVAL CALCULATIONS
Based on 10 bootstrap replicates

CALL :
boot.ci(boot.out = test.result, type = "perc", index = 8)

Intervals :
Level      Percentile
95%    (-0.1044,  0.0624 )
BOOTSTRAP CONFIDENCE INTERVAL CALCULATIONS
Based on 100 bootstrap replicates

boot.ci(test.result, type="perc", index=9) #SCHOOL
VARIANCE#
```

```
BOOTSTRAP CONFIDENCE INTERVAL CALCULATIONS
Based on 100 bootstrap replicates

CALL :
boot.ci(boot.out = test.result, type = "perc", index = 9)

Intervals :
Level      Percentile
95%     ( 0.2933,   0.4523 )
BOOTSTRAP CONFIDENCE INTERVAL CALCULATIONS
Based on 100 bootstrap replicates
```

On the basis of these results, we would conclude that the intercept was significantly different from 0, as were the coefficients for numsense, L _ Free, ISTEPM, and specialed. These results are similar to those for the standard model (model8.5), with the exception of for salary. In addition, the school variance was also significantly different from 0, indicating that there were differences among the schools on the response variables.

Finally, we can fit the model with random coefficients for numsense and ITESPM using the lasso estimator as follows:

```
model8.8 <- glmmLasso(score2~numsense+L_Free+L_Reduced+Salary+I
STEPM+specialed+female, rnd = list(school.f=~numsense+ISTEPM),
lambda=1, family=binomial(link="logit"), data = mathfinal.nomiss)

summary(model8.8)

Call:
glmmLasso(fix = score2 ~ numsense + L_Free + L_Reduced + Salary +
    ISTEPM + specialed + female, rnd = list(school.f =
~numsense +
    ISTEPM), data = mathfinal.nomiss, lambda = 1, family =
binomial(link = "logit"))

Fixed Effects:

Coefficients:
              Estimate StdErr z.value p.value
(Intercept) -1.1170e+01    NA      NA      NA
numsense     6.1897e-02    NA      NA      NA
L_Free      -4.9322e-03    NA      NA      NA
L_Reduced    0.0000e+00    NA      NA      NA
Salary      -2.7014e-05    NA      NA      NA
ISTEPM       1.4005e-01    NA      NA      NA
specialed   -9.5464e-01    NA      NA      NA
female      -3.7284e-02    NA      NA      NA

Random Effects:

StdDev:
                school.f school.f:numsense school.f:ISTEPM
school.f      0.06873047       0.0156410491     0.0161568017
```

```
school.f:numsense 0.01564105     0.0136447387    -0.0009154603
school.f:ISTEPM    0.01615680    -0.0009154603     0.0771196527
```

The request for random coefficients comes from `rnd=list(school.f=~numsense+ISTEPM)`.

The standard deviation estimates are 0.07 for the random school effect, 0.01 for the random `numsense` coefficient, and 0.08 for the random `ISTEPM` coefficient. These values are reduced when compared to those from the standard estimator approach.

MGLM for an Ordinal Outcome Variable

As was the case for non-multilevel data, the cumulative logits link function can be used with ordinal data in the context of multilevel logistic regression, using the cumulative logit link. Furthermore, the multilevel aspects of the model, including random intercept and coefficient take the same form as what we described earlier. The link function and basic model format is very similar to the cumulative logits model described in Chapter 4, with the addition of the multilevel structure described earlier. To provide context to this problem, let's again consider the math achievement results for students. In this case, the outcome variable takes one of three possible values for each member of the sample: 1=Failure, 2=Pass, 3=Pass with distinction. In this case, the question of most interest to the researcher is whether a computation aptitude score is a good predictor of status on the math achievement test.

Random Intercept Logistic Regression

In order to fit a multilevel cumulative logits model using R, we install the `ordinal` package, which allows for fitting a variety of mixed-effects models for categorical outcomes. Within this package, the `clmm` function is used to actually fit the multilevel cumulative logits model. Model parameter estimation is achieved using maximum likelihood based on the Newton-Raphson method. Once we have installed this package, we will use the `library(ordinal)` statement to load it. The R command to then fit the model, obtain the results, and the results themselves, appear below. We should also note that the dependent variable needs to be a factor, leading to our use of `as.factor(score)` in the command sequence. It is important to state at this point that currently, there is not an R package available to fit a random coefficients model for the cumulative logits model. Finally, the current example is too large for application of the glmmLasso function in R. Therefore, we will first take a random sample of the data using the sample _ n function from the dplyr library. The script for this example, which can be found at www.routledge.com/9780367408787, includes the R code for randomly sampling individuals for this example. Note that when you replicate this work, your sample will be different from the one used to create this example, and thus the results will differ to some extent.

```
model8.9<-clmm(as.factor(score)~numsense+L_Free+L_Reduced+
Salary+ISTEPM+specialed+female+(1|school), data=mathfinal.
nomiss.sample)

summary(model8.9)

Cumulative Link Mixed Model fitted with the Laplace
approximation

formula: as.factor(score) ~ numsense + L_Free + L_Reduced +
Salary + ISTEPM +
    specialed + female + (1 | school)
data:     mathfinal.nomiss.sample

 link  threshold nobs logLik  AIC      niter      max.grad cond.H
 logit flexible  1000 -749.32 1518.63 254(907)  8.61e+03 1.1e+11

Random effects:
 Groups Name          Variance Std.Dev.
 school (Intercept) 1          1
Number of groups:   school 32

Coefficients:
            Estimate Std. Error z value Pr(>|z|)
numsense    5.696e-02 3.876e-03  14.697    <2e-16 ***
L_Free     -1.123e-02 7.373e-03  -1.523    0.128
L_Reduced   1.149e-02 4.922e-02   0.233    0.815
Salary     -3.158e-05 1.952e-05  -1.618    0.106
ISTEPM      9.255e-02 1.811e-01   0.511    0.609
specialed   2.747e-02 3.714e-01   0.074    0.941
female      1.383e-01 1.379e-01   1.003    0.316
---
Signif. codes: 0 '***' 0.001 '**' 0.01 '*' 0.05 '.' 0.1 ' ' 1

Threshold coefficients:
     Estimate Std. Error z value
1|2     9.763      1.168   8.357
2|3    13.153      1.212  10.850
```

An examination of the results presented earlier reveals that the variance and standard deviation of intercepts across schools are both 1. Given that the variation is not near 0, we would conclude that there appear to be differences in intercepts from one school to the next. In addition, we see that there were no statistically significant relationships between individual predictor variables and the math achievement test. We also obtain estimates of the model intercepts, which are termed thresholds by clmm. As was the case for the single level cumulative logits model, the intercept represents the log odds of the likelihood of one response versus the other (e.g., 1 versus 2) when the value of the predictor variables are all 0. In many (perhaps most) instances, it would be unlikely, or even impossible, for all values of the independent variables to be 0. Therefore, quite often we don't focus our attention on the thresholds, but rather we pay primary attention to the coefficients.

We can fit the same model using the lasso estimator with the glmmLasso function in R. The function call is essentially the same as for dichotomous logistic regression, with the difference being in the distributional family. For the ordinal logistic regression, we will use family=cumulative(), as below. Also note that in this example, we will set $\lambda=1$. Just as with the dichotomous logistic regression problem, in practice, we would try several values of the tuning parameter, compare model fit using AIC and BIC, and then select the one that yields the optimal value. However, due to space considerations, we only provide an example with one λ value.

```
model8.10 <- glmmLasso(score~numsense+L_Free+L_Reduced+Salary
+ISTEPM+specialed+female, rnd = list(school.f=~1), lambda=1,
family=cumulative(), data = mathfinal.nomiss.sample)

summary(model8.10)

Call:
glmmLasso(fix = score ~ numsense + L_Free + L_Reduced + Salary +
     ISTEPM + specialed + female, rnd = list(school.f = ~1),
data = mathfinal.nomiss.sample,
     lambda = 1, family = cumulative())

Fixed Effects:

Coefficients:
             Estimate StdErr z.value p.value
theta1        5.1230e+00     NA      NA      NA
theta2        8.2461e+00     NA      NA      NA
numsense     -3.7395e-02     NA      NA      NA
L_Free        1.6024e-02     NA      NA      NA
L_Reduced     7.6839e-05     NA      NA      NA
Salary        3.2799e-05     NA      NA      NA
ISTEPM       -9.0803e-03     NA      NA      NA
specialed    -1.2357e-03     NA      NA      NA
female       -2.8099e-02     NA      NA      NA

Random Effects:

StdDev:
            school.f
school.f 0.3857084
```

```
model8.10$aic
          [,1]
[1,]  1543.873
> model8.10$bic
          [,1]
[1,]  1667.679
```

None of the model parameters were fully regularized; i.e., they all had non-zero values. Most, though not all of the Lasso estimated coefficients were smaller than those from the standard estimation approach. The standard deviation estimate for school effect was approximately 0.39, which is similar in value to that obtained in the dichotomous logistic regression approach, unlike was the case for the REML (restricted maximum likelihood) estimator. Indeed, of particular note, is that there do not appear to have been convergence problems with the regularized estimation approach, whereas for the maximum likelihood estimator we did have such difficulties. Therefore, we should have more confidence in the Lasso estimates than those from the standard approach.

Multilevel Count Regression Model

To this point, we have been focused on outcome variables of a categorical nature, such as whether an individual achieves a passing score on a math test. Another type of data that does not fit nicely into the standard models assuming normally distributed errors involves counts or rates of some outcome, particularly of rare events. Such variables often follow the Poisson distribution, a major property of which is that the mean is equal to the variance. It is clear that if the outcome variable is a count, its lower bound must be 0; i.e., one cannot have negative counts. This presents a problem to researchers applying the standard linear regression model, as it may produce predicted values of the outcome that are less than 0, and thus are nonsensical. In order to deal with this potential difficulty, Poisson regression can be used by the data analyst. This approach to dealing with count data rests upon the application of the log to the outcome variable, thereby overcoming the problem of negative predicted counts, since the log of the outcome can take any real number value. Thus, when dealing with the Poisson distribution in the form of counts, we will use the log as the link function in fitting the Poisson regression model:

$$\ln(Y) = \beta_0 + \beta_1 x \tag{8.13}$$

In all other respects, the Poisson model is similar to other regression models in that the relationship between the independent and dependent variables is expressed through the slope, β_1. And again, the assumption underlying the Poisson model is that the mean is equal to the variance. This assumption is typically expressed by stating that the overdispersion parameter, $\phi = 1$. The ϕ parameter appears in the Poisson distribution density and thus is a key component in the fitting function used to determine the optimal model parameter estimates in maximum likelihood. A thorough review of this fitting function is beyond the scope of this book. Interested readers are referred to Agresti (2013) for a complete presentation of this issue.

MGLM for Count Data

When working with count data, we are interested in statistical models designed for use with outcome variables that represented the frequency of some event occurring. Typically these events are relatively rare, such as the number of babies in a family. Perhaps the most common distribution associated with such counts is the Poisson, a distribution in which the mean and the variance are equal. It is a fairly straightforward matter to extend the Poisson regression model to the multilevel context, both conceptually and using R with the appropriate packages. In the following sections, we will demonstrate analysis of multilevel count data outcomes in the context of Poisson regression in R. The example to be used involves the number of cardiac warning incidents (e.g., chest pain, shortness of breath, dizzy spells) for 1000 patients associated with 110 cardiac rehabilitation facilities in a large state over a 6-month period. Patients who had recently suffered from a heart attack and who were entering rehabilitation agreed to be randomly assigned to either a new exercise treatment program or the standard treatment protocol. Of particular interest to the researcher heading up this study is the relationship between treatment condition and the number of cardiac warning incidents. The new approach to rehabilitation is expected to result in fewer such incidents as compared to the traditional method. In addition, the researcher has also collected data on the sex of the patients, and the number of hours that each rehabilitation facility is open during the week. This latter variable is of interest as it reflects the overall availability of the rehabilitation programs. The new method of conducting cardiac rehabilitation is coded in the data as 1, while the standard approach is coded as 0. Males are also coded as 1 while females are assigned a value of 0.

Random Intercept Poisson Regression

The R commands and resultant output for fitting the Poisson regression model to the data appear below using the `glmer` function in the `lme4` library that was employed earlier to fit the dichotomous logistic regression models.

```
model8.11<-glmer(heart~trt+sex+(1|rehab),family=poisson,
data=rehab_data)

summary(model8.11)

Generalized linear mixed model fit by maximum likelihood
(Laplace Approximation) ['glmerMod']
 Family: poisson  ( log )
Formula: heart ~ trt + sex + (1 | rehab)
   Data: rehab_data

     AIC      BIC   logLik deviance df.resid
 11470.2  11489.8  -5731.1  11462.2      996

Scaled residuals:
   Min      1Q Median      3Q     Max
-5.906 -1.695 -0.881   0.756 40.163

Random effects:
 Groups Name        Variance Std.Dev.
 rehab  (Intercept) 1.216    1.103
Number of obs: 1000, groups:  rehab, 110

Fixed effects:
            Estimate Std. Error z value Pr(>|z|)
(Intercept)  0.83408    0.11181    7.46 8.67e-14 ***
trt         -0.45612    0.03389  -13.46  < 2e-16 ***
sex          0.39305    0.03344   11.76  < 2e-16 ***
---
Signif. codes:  0 '***' 0.001 '**' 0.01 '*' 0.05 '.' 0.1 ' ' 1

Correlation of Fixed Effects:
    (Intr) trt
trt -0.112
sex -0.167 -0.055
```

In terms of the function call, the syntax for Model 8.11 is virtually identical to that used for the dichotomous logistic regression model, with the major difference being that `family=poisson`. The dependent and independent variables are linked in the usual way that we have seen in R: `heart~trt+sex`. Here, the outcome variable is `heart`, which reflects the frequency of the warning signs for heart problems that we described earlier. The independent variables are treatment (`trt`) and `sex` of the individual, while the specific rehabilitation facility is contained in the variable `rehab`. In this model, we are fitting a random intercept only, with no random slope and no rehabilitation center level variables.

The results of the analysis indicate that there is variation among the intercepts from rehabilitation facility to rehabilitation facility, with a variance of 1.216. As a reminder, the intercept reflects the mean frequency of events when (in this case) both of the independent variables are 0; i.e., females in the control condition. The average intercept across the 110 rehabilitation centers is 1.216, and this non-zero value suggests that the intercept does differ from center to center. Put another way, we can conclude that the mean number of cardiac warning signs varies across rehabilitation centers and that the average female in the control condition will have approximately 1.2 such incidents over the course of 6 months. In addition, these results reveal a statistically significant negative relationship between heart and trt, and a statistically significant positive relationship between heart and sex. Remember that the new treatment is coded as 1 and the control as 0, so that a negative relationship indicates that there are fewer warning signs over 6 months for those in the treatment than those in the control group. Also, given that males were coded as 1 and females as 0, the positive slope for sex means that males have more warning signs on average than do females.

Random Coefficient Poisson Regression

If we believe that the treatment will have different impacts on the number of warning signs present among the rehabilitation centers, we would want to fit the random coefficient model. This can be done for Poisson regression just as it was syntactically for dichotomous logistic regression, as demonstrated in Model 8.11.

```
model8.12<-glmer(heart~trt+sex+(trt|rehab),family=poisson,
data=rehab_data)

summary(model8.12)

Generalized linear mixed model fit by maximum likelihood
(Laplace Approximation) ['glmerMod']
 Family: poisson  ( log )
Formula: heart ~ trt + sex + (trt | rehab)
   Data: rehab_data

     AIC      BIC   logLik deviance df.resid
 10554.6  10584.0  -5271.3  10542.6      994

Scaled residuals:
   Min     1Q Median     3Q    Max
-6.109 -1.552 -0.725  0.640 31.917

Random effects:
 Groups Name        Variance Std.Dev. Corr
 rehab  (Intercept) 1.869    1.367
```

```
          trt              1.844      1.358      -0.62
Number of obs: 1000, groups:   rehab, 110

Fixed effects:
            Estimate Std. Error z value Pr(>|z|)
(Intercept)  0.52852    0.14124    3.742 0.000183 ***
trt         -0.12222    0.14749   -0.829 0.407310
sex          0.34415    0.03523    9.769  < 2e-16 ***
---
Signif. codes:  0 '***' 0.001 '**' 0.01 '*' 0.05 '.' 0.1 ' ' 1

Correlation of Fixed Effects:
    (Intr) trt
trt -0.622
sex -0.137 -0.004
```

The syntax for the inclusion of random slopes in the model is identical to that used with logistic regression and thus will not be commented on further here. The random effect for slopes across rehabilitation centers was estimated to be 1.844, indicating that there is some differential center effect to the impact of treatment on the number of cardiac warning signs experienced by patients. Indeed, the variance for the random slopes is approximately the same magnitude as the variance for the random intercepts, indicating that these two random effects are quite comparable in magnitude. The correlation of the random slope and intercept model components is fairly large and negative (−0.62), meaning that the greater the number of cardiac events in a rehab center, the lower the impact of the treatment on the number of such events. The average slope for treatment across centers was no longer statistically significant, indicating that when we account for the random coefficient effect for treatment, the treatment effect itself goes away.

As with the logistic regression, we can compare the fit of the two models using both information indices, and a likelihood ratio test.

```
anova(model8.11,model8.12)
Data: rehab_data
Models:
model8.11: heart ~ trt + sex + (1 | rehab)
model8.12: heart ~ trt + sex + (trt | rehab)
        Df   AIC   BIC  logLik deviance  Chisq Chi Df
Pr(>Chisq)
model8.11  4 11470 11490 -5731.1    11462
model8.12  6 10555 10584 -5271.3    10543 919.63      2 <
2.2e-16 ***
---
Signif. codes:  0 '***' 0.001 '**' 0.01 '*' 0.05 '.' 0.1 ' ' 1
```

Given that there is a statistically significant difference in model fit (p<0.001), and Model 8.8 has the smaller AIC and BIC values, these results provide further statistical evidence that the relationship of treatment with the number of cardiac symptoms differs across rehabilitation centers.

We can fit a regularized Poisson regression model using glmmLasso, with the function call taking generally the same form as in the previous examples in this chapter. Given the count nature of the data we will set the family equal to poisson. Also, as with the logistic regression examples, we first need to create a factor version of the clustering variable rehab. We will then fit the model and obtain the summary with the following set of commands:

```
rehab_data<-read.spss("c:\\research\\regularization book\\
data\\rehab chapter 8.sav", to.data.frame=TRUE, use.value.
labels=FALSE)

rehab_data$rehab.f<-as.factor(rehab_data$rehab)

model8.13 <- glmmLasso(heart~trt+sex+hours, rnd =
list(rehab.f=~1), lambda=1, family=poisson(), data = rehab_
data)

summary(model8.13)

Call:
glmmLasso(fix = heart ~ trt + sex + hours, rnd =
list(rehab.f = ~1),
    data = rehab_data, lambda = 1, family = poisson())

Fixed Effects:

Coefficients:
            Estimate StdErr z.value p.value
(Intercept)  0.97672    NA     NA      NA
trt         -0.45416    NA     NA      NA
sex          0.38717    NA     NA      NA
hours        0.22374    NA     NA      NA

Random Effects:

StdDev:
          rehab.f
rehab.f 0.8223291
```

The coefficient estimates from the lasso approach were very similar to those from the standard restricted maximum likelihood estimator (REML). However, the standard deviation from the random effect was shrunken by the Lasso estimator when compared to the standard approach. We can use the bootstrap to obtain confidence intervals for the parameter estimates from the glmmLasso. The script is available on the book website www.routledge.com/9780367408787.

```
test.result<-boot(data=rehab_data, statistic=glmmlasso.
poisson.boot, R=100)

boot.ci(test.result, type="perc", index=1)  #INTERCEPT#
```

```
BOOTSTRAP CONFIDENCE INTERVAL CALCULATIONS
Based on 100 bootstrap replicates

CALL :
boot.ci(boot.out = test.result, type = "perc", index = 1)

Intervals :
Level     Percentile
95%    ( 0.7127,  0.9754 )
Calculations and Intervals on Original Scale
boot.ci(test.result, type="perc", index=2) #TRT#
BOOTSTRAP CONFIDENCE INTERVAL CALCULATIONS
Based on 100 bootstrap replicates

CALL :
boot.ci(boot.out = test.result, type = "perc", index = 2)

Intervals :
Level     Percentile
95%    (-0.6429, -0.3581 )
Calculations and Intervals on Original Scale
boot.ci(test.result, type="perc", index=3) #SEX#
BOOTSTRAP CONFIDENCE INTERVAL CALCULATIONS
Based on 100 bootstrap replicates

CALL :
boot.ci(boot.out = test.result, type = "perc", index = 3)

Intervals :
Level     Percentile
95%    ( 0.0622,  0.4800 )
Calculations and Intervals on Original Scale

boot.ci(test.result, type="perc", index=4) #HOURS#
BOOTSTRAP CONFIDENCE INTERVAL CALCULATIONS
Based on 100 bootstrap replicates

CALL :
boot.ci(boot.out = test.result, type = "perc", index = 4)

Intervals :
Level     Percentile
95%    ( 0.0154,  0.2935 )
Calculations and Intervals on Original Scale

boot.ci(test.result, type="perc", index=5) #REHAB INTERCEPT
VARIANCE#
BOOTSTRAP CONFIDENCE INTERVAL CALCULATIONS
Based on 100 bootstrap replicates

CALL :
boot.ci(boot.out = test.result, type = "perc", index = 5)

Intervals :
Level     Percentile
```

```
95%     ( 0.9012,  1.1643 )
Calculations and Intervals on Original Scale
```

On the basis of these results, we will conclude that all of the parameter estimates likely differ from 0 in the population. This result matches what we obtained from the standard estimator.

As with the REML estimator, we can fit a random coefficients Poisson regression model using glmmLasso. The basic structure is similar to that for logistic regression, where the variable for which we want the random coefficient estimate appears as rnd = list(rehab.f=~trt).

```
model8.14 <- glmmLasso(heart~trt+sex+hours, rnd = list(rehab.
f=~trt), lambda=1, family=poisson(), data = rehab_data)

summary(model8.14)

Call:
glmmLasso(fix = heart ~ trt + sex + hours, rnd =
list(rehab.f = ~trt),
    data = rehab_data, lambda = 1, family = poisson())

Fixed Effects:

Coefficients:
              Estimate StdErr z.value p.value
(Intercept)   0.81322     NA     NA      NA
trt          -0.16028     NA     NA      NA
sex           0.34271     NA     NA      NA
hours         0.17634     NA     NA      NA

Random Effects:

StdDev:
                  rehab.f  rehab.f:trt
rehab.f         0.52241450 -0.07295662
rehab.f:trt    -0.07295662  0.25911672
```

Once again, the coefficients for the fixed effects are similar for the two (REML and Lasso) estimators. However, for the random effects, the Lasso approach greatly shrank the variances. In particular, the REML estimator yielded standard deviations for the intercept and coefficient of 1.345 and 1.352, respectively, whereas Lasso yielded 0.52 and 0.26, respectively. The correlation between the random effects for the REML estimator was −0.62, whereas for Lasso it was −0.07, once again reflecting the impact of the penalty on the estimation.

We can use the bootstrap for the purpose of inference.

```
test.result<-boot(data=rehab_data, statistic=glmmlasso.
poisson.boot, R=100)
boot.ci(test.result, type="perc", index=1) #INTERCEPT#
BOOTSTRAP CONFIDENCE INTERVAL CALCULATIONS
```

```
Based on 100 bootstrap replicates

CALL :
boot.ci(boot.out = test.result, type = "perc", index = 1)

Intervals :
Level      Percentile
95%    ( 0.5897,  1.2245 )
Calculations and Intervals on Original Scale
boot.ci(test.result, type="perc", index=2) #TRT#
BOOTSTRAP CONFIDENCE INTERVAL CALCULATIONS
Based on 100 bootstrap replicates

CALL :
boot.ci(boot.out = test.result, type = "perc", index = 2)

Intervals :
Level      Percentile
95%    (-0.2787,  0.0765 )
Calculations and Intervals on Original Scale
boot.ci(test.result, type="perc", index=3) #SEX#
BOOTSTRAP CONFIDENCE INTERVAL CALCULATIONS
Based on 100 bootstrap replicates

CALL :
boot.ci(boot.out = test.result, type = "perc", index = 3)

Intervals :
Level      Percentile
95%    (-0.1476,  0.3936 )
Calculations and Intervals on Original Scale
boot.ci(test.result, type="perc", index=4) #HOURS#
BOOTSTRAP CONFIDENCE INTERVAL CALCULATIONS
Based on 100 bootstrap replicates

CALL :
boot.ci(boot.out = test.result, type = "perc", index = 4)

Intervals :
Level      Percentile
95%    ( 0.0000,  0.1648 )
Calculations and Intervals on Original Scale
boot.ci(test.result, type="perc", index=5) #REHAB INTERCEPT
VARIANCE#
BOOTSTRAP CONFIDENCE INTERVAL CALCULATIONS
Based on 100 bootstrap replicates

CALL :
boot.ci(boot.out = test.result, type = "perc", index = 5)

Intervals :
Level      Percentile
95%    ( 0.0938,  0.8797 )
Calculations and Intervals on Original Scale
```

```
boot.ci(test.result, type="perc", index=6) #TRT SLOPE VARIANCE#
BOOTSTRAP CONFIDENCE INTERVAL CALCULATIONS
Based on 100 bootstrap replicates

CALL :
boot.ci(boot.out = test.result, type = "perc", index = 6)

Intervals :
Level      Percentile
95%    ( 0.0886,  0.5625 )
Calculations and Intervals on Original Scale
boot.ci(test.result, type="perc", index=7) #RANDOM EFFECT
CORRELATION#
BOOTSTRAP CONFIDENCE INTERVAL CALCULATIONS
Based on 100 bootstrap replicates

CALL :
boot.ci(boot.out = test.result, type = "perc", index = 7)

Intervals :
Level      Percentile
95%    (-0.0992,  0.1623 )
Calculations and Intervals on Original Scale
```

These confidence interval results suggest that the intercept and the variances for the random intercept and random slope were significantly different from 0; i.e., for these terms 0 was not in the 95% confidence intervals. These results differ with respect to the sex effect but otherwise yield similar qualitative conclusions.

Finally, we need to ascertain which model yields the better fit to the data. Recall that for the REML estimator, the random trt coefficients model yielded a statistically significantly better fit, and both AIC and BIC were smaller for this model as well. For the Lasso estimated models, the AIC and BIC values appear as follows:

```
model8.13$aic
          [,1]
[1,] -7442.299
model8.13$bic
          [,1]
[1,]  -6935.797

model8.14$aic
          [,1]
[1,] -8300.554
model8.14$bic
          [,1]
[1,]  -7391.146
```

The AIC and BIC were lower for the random intercept and slope model, as compared to the random intercept only. Thus, as for the REML estimator, we would select the random intercept and slope model as being optimal.

Summary

Our focus in this chapter was on models designed for use situations in which individuals are nested together within clusters (e.g., students within schools). Standard statistical models are not appropriate in such cases because they don't take account of the correlations among members of the sample who are grouped within the same cluster. Ignoring this nesting can lead to biased parameter and standard error estimates. However, multilevel models, which take account of this structure, can be used to unbiased parameter estimates for a wide variety of dependent variable types. As we saw, common models such as regression, logistic regression, and Poisson regression, can all be extended for use in the multilevel context. Furthermore, we saw that regularized estimators can also be easily applied to multilevel models such as these, using functions in the R software package. To aid with inference, we have written a bootstrap sampling script that is available on the website for this book, www.routledge.com/9780367408787.

The general principles are applicable in the context of multilevel modeling, just as they were for the other model types described in this chapter. We will need to identify the optimal value of the regularization tuning parameter for a given model, typically using information indices such as AIC and BIC. In the multilevel context, there is also the added complexity associated with decisions around whether to include a random intercept only term or also a random coefficient. This is not a concern when using single-level models, and thus adds an additional layer of decision-making to go along with determining the optimal tuning parameter value. However, as we saw in this chapter such decisions can be made using information indices.

References

Agresti, A. (2013). *Categorical data analysis*. Hoboken, NJ: John Wiley & Sons.

Akaike, H. (1973). Information theory and an extension of the maximum likelihood principle. In B. N. Petrov & F. Csaki (Eds.), *International symposium on information theory* (pp. 267–281).

Bandalos, D. L., & Finney, S. J. (2019). Factor analysis: Exploratory and confirmatory. In G. R. Hancock, L. M. Stapleton, & R. O. Mueller (Eds.), *The reviewer's guide to quantitative methods in the social sciences*. New York: Routledge.

Bauer, D. J., & Hussong, A. M. (2009). Psychometric approaches for developing commensurate measures across independent studies: Traditional and new models. *Psychological Methods, 14*(2), 101–125.

Bollen, K. A. (1989). *Structural equations with latent variables*. New York: John Wiley & Sons.

Bollen, K. A. (1990). Overall fit in covariance structure models: Two types of sample size effects. *Psychological Bulletin, 107*(2), 256–259.

Bollen, K. A. (1996). An alternative two stage least squares (2SLS) estimator for latent variable equations. *Psychometrika, 61*(1), 109–121.

Bollen, K. A., & Bauer, D. J. (2004). Automating the selection of model-implied instrumental variables. *Sociological Methods & Research, 32*(4), 425–452.

Bollen, K. A., & Stine, R. A. (1992). Bootstrapping goodness-of-fit measures in structural equation models. *Sociological Methods & Research, 21*(2), 205–229.

Brown, T. A. (2015). *Confirmatory factor analysis for applied research*. New York: The Guilford Press.

Bühlmann, P., & van de Geer, S. (2011). *Statistics for high-dimensional data*. New York: Springer.

Caron, P.-O. (2018). Minimum average partial correlation and parallel analysis: The influence of oblique structures. *Communications in Statistics—Simulation and Computation, 48*(7), 2110–2117.

Clark, D. A., & Bowles, R. P. (2018). Model fit and item factor analysis: Overfactoring, underfactoring, and a program to guide interpretation. *Multivariate Behavioral Research, 53*, 544–558.

Cohen, J. (1988). *Statistical power analysis for the behavioral sciences*. New York: Lawrence Erlbaum Associates, Publishers.

Curran, P. J., West, S., & Finch, J. F. (1996). The robustness of test statistics to non-normality and specification error in confirmatory factor analysis. *Psychological Methods, 1*(1), 16–29.

Dudoit, S., & Fridlyand, J. (2002). A prediction-based resampling method for estimating the number of clusters in a dataset. *Genome Biology, 3*(7), 1–21.

Efron, B. (1982). *The jackknife, the bootstrap and other resampling plans*. CBMS-NSF Regional Conference Series in Applied Mathematics, Monograph 38, SIAM, Philadelphia.

Fabrigar, L. R., & Wegener, D. T. (2012). *Exploratory factor analysis*. Oxford: Oxford University Press.

Finch, W. H. (2020). Using fit statistic differences to determine the optimal number of factors to retain in an exploratory factor analysis. *Educational and Psychological Measurement, 80*(2), 217–241.

Finch, W. H., Hernandez Finch, M. E., & Moss, L. (2014). Dimension reduction regression techniques for high dimensional data. *Multiple Linear Regression Viewpoints, 40*(2), 1–15.

Finney, S. J., & DiStefano, C. (2013). Nonnormal and categorical data in structural equation modeling. In G. R. Hancock & R. O. Mueller (Eds.), *Structural equation modeling: A second course* (pp. 439–492). Charlotte, NC: Information Age Publishers.

Flora, D. B., & Curran, P. J. (2004). An empirical evaluation of alternative methods of estimation for confirmatory factor analysis with ordinal data. *Psychological Methods, 9*(4), 466–491.

Fox, J. (2016). *Applied regression analysis & generalized linear models.* Thousand Oaks, CA: Sage.

Friedman, J. H., Hastie, T., & Tibshirani, R. (2010). Regularization paths for generalized linear models via coordinate descent. *Journal of Statistical Software, 33*(1), 1–22.

Garrido, L. E., Abad, F. J., & Ponsoda, V. (2011). Performance of velicer's minimum average partial factor retention method with categorical variables. *Educational and Psychological Measurement, 71*(3), 551–570.

Gelman, A., Hill, J., & Vehtari, A. (2021). *Regression and other stories.* Cambridge: Cambridge University Press.

Geminiani, E., Marra, G., & Moustaki, I. (2021). Single and multiple group penalized factor analysis: A trust-region algorithm approach with integrated automatic multiple tuning parameter selection. *Psychometrika, 86*(1), 65–95.

Gorsuch, R. (1983). *Factor analysis,* 2nd ed. Hillsdale, NJ: Lawrence Erlbaum Associates.

Hahs-Vaughn, D. L. (2017). *Applied multivariate statistical concepts.* New York: Routledge Taylor and Francis Group.

Hastie, T., Tibshirani, R., & Friedman, J. (2011). *The elements of statistical learning: Data mining, inference, and prediction.* New York: Springer.

Hastie, T., Tibshirani, R., & Wainwright, M. (2015). *Statistical learning with sparsity: The lasso and generalizations.* Boca Raton, FL: CRC Press Taylor & Francis group.

Hirose, K., & Yamamoto, M. (2015). Sparse estimation via nonconcave penalized likelihood in factor analysis model. *Statistical Computing, 25,* 863–875.

Hoerl, A. E., & Kennard, R. W. (1970). Ridge regression: Biased estimation for nonorthogonal problems. *Technometrics, 12*(1), 55–67.

Horn, J. L. (1965). A rationale and test for the number of factors in factor analysis. *Psychometrika, 30,* 179–185.

Hu, L.-T., & Bentler, P. M. (1999). Cutoff criteria for fit indexes in covariance structure analysis: Conventional criteria versus new alternatives. *Structural Equation Modeling, 6*(1), 1–55.

Huang, P.-H., Chen, H., & Weng, L.-J. (2017). A penalized likelihood method for structural equation modeling. *Psychometrika, 82,* 329–354.

Jacobucci, R., Grimm, K. J., & Mcardle, J. J. (2016). Regularized structural equation modeling. *Structural Equation Modeling: A Multidisciplinary Journal, 23*(4), 555–566.

Johnston, J. (1984). *Econometric methods,* 3d ed. New York: McGraw-Hill.

Jung, S. (2013). Structural equation modeling with small sample sizes using two-stage ridge least-squares estimation. *Behavioral Research Methods, 45*(1), 75–81.

Kaplan, D. (2014). *Bayesian statistics for the social sciences.* New York: The Guilford Press.

Kline, R. B. (2016). *Principles and practice of structural equation modeling.* New York: The Guilford Press.

Liu, Q., & Pierce, D. A. (1994). A note on Gauss-Hermite quadrature. *Biometrika, 81*(3), 624–629.

Lockhart, R., Taylor, J., Tibshirani, R., & Tibshirani, R. (2014). A significance test for the lasso. *Annals of Statistics, 42*(2), 413–468.

Meinhausen, N., & Bühlmann, P. (2010). Stability selection. *Journal of the Royal Statistical Society, B, 72*(4), 417–473.

Muthén, B., du Toit, S. H. C., & Spisic, D. (1997). *Robust inference using weighted least squares and quadratic estimating equations in latent variable modeling with categorical and continuous outcomes.* Unpublished technical report.

Park, T., & Casella, G. (2008). The Bayesian Lasso. *Journal of the American Statistical Association, 103*(482), 681–686.

Raiche, G., Walls, T.A., Magis, D., Riopel, M., & Blais, J.-G. (2012). Non-graphical solutions for Cattell's scree test. *Methodology, 9*(1), 23–29.

Revelle, W., & Rocklin, T. (1979). Very simple structure: An alternative procedure for estimating the optimal number of interpretable factors. *Multivariate Behavioral Research, 14,* 403–414.

Rousseeuw, P. J. (1987). Silhouettes: A graphical aid to the interpretation and validation of cluster analysis. *Computational and Applied Mathematics, 20,* 53–65.

Ruscio, J., & Roche, B. (2012). Determining the number of factors to retain in an exploratory factor analysis using comparison data of known factorial structure. *Psychological Assessment, 24,* 282–292.

Satorra, A., & Bentler, P. M. (1994). Corrections to test statistics and standard errors in covariance structure analysis. In A. von Eye & C. C. Clogg (Eds.), *Latent variables analysis: Applications for developmental research* (pp. 399–419). Thousand Oaks, CA: Sage Publications, Inc.

Schelldorfer, J., Bühlmann, P., & van de Geer, S. (2011). Estimation for high-dimensional linear mixed-effects models using l1-penalization. *Scandinavian Journal of Statistics, 38*(2), 197–214.

Schwarz, G. (1978). Estimating the dimension of a model. *Annals of Statistics, 6,* 461–464.

Tabachnick, B. G., & Fidell, L. S. (2019). *Using multivariate statistics.* New York: Pearson.

Thompson, B. (2004). *Exploratory and confirmatory factor analysis: Understanding concepts and applications.* Washington, DC: American Psychological Association.

Thurstone, L. L. (1947). *Multiple factor analysis.* Chicago: Chicago Press.

Tibshirani, R. (1996). Regression shrinkage and selection via the Lasso. *Journal of the Royal Statistical Society, B, 58*(1), 267–288.

Tibshirani, R., Walther, G., & Hastie, T. (2001). Estimating the number of clusters in a data set via the gap statistic. *Journal of the Royal Statistical Society, B, 63*(2), 411–423.

Tofighi, D., & Enders, C. K. (2007). Identifying the correct number of classes in a growth mixture model. In G. R. Hancock (Ed.), *Advances in latent variable mixture models* (pp. 317–341). Greenwich: Information Age.

Velicer, W. F. (1976). Determining the number of components from the matrix of partial correlations. *Psychometrika, 41*(3), 321–327.

Wirth, R. J., & Edwards, M. C. (2007). Item factor analysis: Current approaches and future directions. *Psychological Methods, 12*(1), 58–79.

Witten, D. M., & Tibshirani, R. (2010). A framework for feature selection in clustering. *Journal of the American Statistical Association, 105*(490), 713–726.

Yuan, K., & Bentler, P. (1997). Mean and covariance structure analysis: Theoretical and practical improvements. *Journal of the American Statistical Association, 92,* 767–774.

Yuan, K.-H., Bentler, P. M., & Kano, Y. (1997). On averaging variables in a confirmatory factor analysis model. *Behaviormetrika, 24*(1), 71–83.

Yuan, K.-H., Bentler, P. M., & Zhang, W. (2005). The effect of skewness and kurtosis on mean and covariance structure analysis: The univariate case and its multivariate implication. *Sociological Methods & Research, 34*(2), 240–258.

Yuan, M., & Lin, Y. (2006). Model selection and estimation in regression with grouped variables. *Journal of the Royal Statistical Society, B, 68*(1), 49–67.

Zhang, C-H. (2010). Nearly unbiased variable selection under minimax concave penalty. *Annals of Statistics, 38*(2), 894–942.

Zhao, P., Rocha, G., & Yu, B. (2009). Grouped and hierarchical model selection through composite absolute penalties. *Annals of Statistics, 37*(6A), 3486–3497.

Zhao, P., & Yu, B. (2007). On model selection consistency of lasso. *Journal of Machine Learning Research, 7*(2), 2541.

Zou, H. (2006). The adaptive lasso and its oracle properties. *Journal of the American Statistical Association, 101*(476), 1418–1429.

Zou, H., & Hastie, T. (2005). Regularization and variable selection via the elastic net. *Journal of the Royal Statistical Society, B, 67*(2), 301–320.

Zwick, W. R., & Velicer, W. F. (1986). Comparison of five rules for determining the number of components to retain. *Psychological Bulletin, 99*(3), 432–442.

Index

Note: Page numbers in *italics* indicate a figure and page numbers in **bold** indicate a table on the corresponding page.

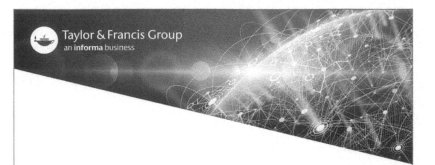

Milton Keynes UK
Ingram Content Group UK Ltd.
UKHW031533071024
449327UK00005B/93

9 780367 408787